The Living Collection

DAVID RAE

The Living Collection

ROYAL BOTANIC
GARDEN EDINBURGH
2011

Royal
Botanic Garden
Edinburgh

Published by the Royal Botanic Garden Edinburgh
20A Inverleith Row, Edinburgh EH3 5LR

ISBN 978–1–906129–77–4

The Royal Botanic Garden Edinburgh (RBGE) confirms that all address details and URLs of websites are accurate at time of going to press. RBGE cannot be responsible for details which become out of date after publication.

The Royal Botanic Garden Edinburgh is a charity registered in Scotland (no. SC007983) and is supported by the Scottish Government Rural and Environment Science and Analytical Services.

Designed and typeset in Minion and Shaker
by Nye Hughes, Dalrymple
Printed by Oriental Press, Dubai, United Arab Emirates
Printed on G Print Matt, manufactured to ISO14001
and EMAS international standards, minimising negative
impacts on the environment

Frontispiece: Sacred lotus (*Nelumbo nucifera*)
Front Cover: *Meconopsis* 'Slieve Donard' (Infertile Blue Group) growing at Dawyck Botanic Garden, one of 65 taxa and 21 species of *Meconopsis* in the Living Collection of the Royal Botanic Garden Edinburgh. *Meconopsis* are a 'signature' genus of the Garden's Collection on several counts – they have been cultivated at the Garden since they were first introduced into cultivation, along with countless other Sino-Himalayan species in the early 20th century, they are found in many of the countries in which the Garden works (e.g. China, Nepal and Bhutan), they grow well in the cool Scottish climate and they have been the subject of a recent in-depth study by the 'Meconopsis Group' who have used the Garden's nursery to grow, study, compare and describe species and cultivars within the genus.

Dedicated to the horticulture staff who have curated and maintained the Living Collection since 1670

FOREWORD

One of Scotland's National Collections is the living plants at the Royal Botanic Garden Edinburgh. With the library, archive and three million herbarium specimens they are the heart of one of the world's leading scientific botanic gardens.

National collections have a profound influence on the nation's psyche. No one has shown this better than Neil MacGregor, whose *A History of the World in 100 Objects* tells the stories of artefacts from humanity's two-million-year journey. It may appear absurd to compare the treasures of the British Museum with a collection of plants that could be grown from cuttings or packets of seeds. But the value of plants is simply different from, not less than, that of our greatest artefacts, and the tale a living collection can tell is just a different one. It is a much longer story, tracing back over 400 million years, with human beings arriving only in the most recent chapter. Like all histories it holds important lessons for the future, a future that depends on plants. After all, humans could only make their entry into the world after plants had first made it habitable by creating a breathable atmosphere and sources of food for ourselves and other animals. Yet the very familiarity of plants, the green background to our daily lives, allows us to overlook the fact that we can't live without them.

In this richly illustrated book, David Rae tells the story of a botanic garden from the perspective of its living collections: their origins, purposes and the many uses they can be put to. Using historical images of the plant collections and the people who have shaped them, he charts its evolution from physic garden to biodiversity research institute and shows how education and conservation have been emerging themes in recent decades. In doing so he opens up and explores the true importance of the living collections and exposes the potential botanic gardens have to preserve the planet's green heritage for the benefit – and perhaps even the survival – of future generations.

STEPHEN BLACKMORE CBE FRSE
Regius Keeper and Queen's Botanist

left A view across the Edinburgh Garden from the Tropical Palm House roof.

As Director of Horticulture at the Royal Botanic Garden Edinburgh (RBGE), my main responsibility is to oversee the curation of the Living Collection of plants which is spread over four gardens. The term 'Living Collection' may seem strange – why wouldn't the collection of plants be living? – but it is used widely in botanic gardens to differentiate these collections from dried or preserved collections which reside in herbaria. The concept of 'curation' has particular resonance as in many respects the plants in botanic gardens can be likened to the collections in a museum or library. Many of the same concepts and processes apply, such as selection, representation and display, recording and meticulous care and maintenance. The library analogy comes to mind when visitors ask me why we grow so many plants from so many places around the world. In this respect the Collection is like

a library: not all of the books are used all of the time; in fact only a small core may be used on a regular basis but others are held in the collection ready and accessible for when they are needed. The analogy reaches further – libraries and museums don't hold random collections of books or artefacts, but specialise in particular areas, aiming to ensure that coverage of those areas is as comprehensive as possible to facilitate specialist research or other uses. The analogies end here, though, because living collections in botanic gardens are just that – alive and growing, rather than inanimate objects, and they demand particular microclimates, soils and cultivation techniques which present their own challenges and delights.

The Living Collection is both a useful resource and also a visitor attraction and it is my responsibility to ensure that we grow the plants needed for all the diverse uses of the Collection, whether advanced systematic or molecular research, conservation programmes, education at all levels or recreation. I am constantly amazed at the number of ways in which the plants are used and it is deeply satisfying trying to help all those who need to use them for their work, or who simply want them as a high-quality background for events or recreation. It is essential that the plants are in good condition and available when needed, in the right quantity and location and that accurate, accessible records are kept. Of equal importance is to make sure that all four of our Gardens are well maintained, beautiful and a pleasure to visit. Working to meet these diverse demands is a constant but invigorating challenge.

I have wanted to write a book for some time to explain why we cultivate the particular selection of plants at RBGE and to describe the extent of the curation work involved in ensuring the Living Collection is 'fit for purpose' for all its users. This book is the result. The first section shows how the plants have been gathered over more than 340 years to create the current Living Collection. I trace the Garden's story through its complex history, involving at least three previous locations before the current site of the Edinburgh Garden, fifteen Regius Keepers, nineteen Directors of Horticulture (or Curators), changing financial structures and management systems and the acquisition of the three Regional Gardens. In

section two I reveal the extensive work which is often 'behind the scenes' and not seen by visitors to the Gardens – how we care for and record the collections. I then describe how the Living Collection is used for research, education and, increasingly, conservation.

In describing the history, management and use of the Living Collection I try to show that it, and the four wonderful Gardens within which it resides, are the core backbone on which all else depends. The plants and gardens, supported by the preserved collections and library, are the foundation on which we base our research, conservation, education, exhibitions and events, and their diversity and quality gives us the authority and credibility to deliver international capacity building and training in more than 40 countries.

To some, a scientifically arranged and meticulously recorded collection of plants might seem a luxury in these financially challenging times. However, RBGE and its Living Collection have always evolved with the times to meet the challenges of the day and, despite more than three centuries of history, this is still true today. The Living Collection is needed now more than ever. It is being used actively right now to support cutting-edge research, to reintroduce threatened plant species back into the wild and to create four beautiful landscapes in which to relax in these increasingly stressful times. It is being used to train the next generation of scientists who will help us to catalogue and better understand our environment and horticulturists who will help us care for our precious world. It creates the Gardens which inspire us culturally through art, exhibitions and live performance and it is a resource for teaching children about nature and its importance.

The Royal Botanic Garden Edinburgh is a wonderful and fascinating place to work. Staff from all departments, including horticulture, science, education, visitor services and administration, work together to create a remarkable institution that has a vital contribution to make to Scotland and the wider world. It is rich in history but with a clear mission for today – 'to explore and explain the world of plants for a better future'.

DAVID RAE
Director of Horticulture, Royal Botanic Garden Edinburgh

A gallery of living plants

The Living Collection in RBGE's four Gardens can be seen as a library, museum or art gallery displaying the extraordinary diversity and visual impact of plants. Their diversity is central to scientific research and conservation but their beauty is probably what attracts most people to spend time in the Garden.

Today's botanical displays are generally grouped into four distinct (though overlapping) categories: taxonomic, geographic, ecological and thematic. Carefully curated specimens are part of an integrated and beautifully displayed whole yet each plant can be enjoyed just as it is: one of nature's 'works of art'.

Rhododendrons

Probably the Garden's most iconic plant group, rhododendrons grow especially well in the mild, moist climate of Benmore but they also thrive in Edinburgh's highly acidic soil despite the cooler, drier environment. Tropical species in section *Vireya*, found in the mountains of South-East Asia and cultivated in Edinburgh's glasshouses, are also a speciality. Today's internationally renowned collection is a legacy of the research which began with the influx of new plants when China opened to the west in the late 19th century.

above One of Benmore's magnificent large-leaved rhododendrons: the lilac-flowered *Rhododendron anthosphaerum* was grown from seed collected by Joseph Rock in Yunnan, south west China in 1948.

Ferns

Ferns are the focus of taxonomic research and practical conservation at RBGE but they also add texture, form and subtlety to the landscape of all four Gardens, especially outside at Benmore and Dawyck, in the Fern House at Edinburgh and in the newly restored Victorian Fernery at Benmore. While their tiny spores and two-stage life cycle can create cultivation complications for some, horticultural staff at Edinburgh have made them a speciality, with over 600 species in the Living Collection.

above Not how you'd expect a fern to look: in Edinburgh's Fern House, this species of *Aglaomorpha* from Indonesia illustrates fascinating diversity of form among ferns. The long threads are spore-bearing fronds.

Taxonomic

Taxonomy – the science of identifying, classifying and naming – is central to the work of botanic gardens. Rhododendrons, ferns, conifers and gingers form important taxonomic groups at RBGE because of long-standing research and conservation programmes. The first three groups make a significant contribution to the outdoor landscapes at all four Gardens and the gingers dominate the Tropical Montane House at Edinburgh.

Conifers

Conifers are of special historic interest at Benmore and Dawyck where successive generations of owners took a interest in the latest collections, at first largely from North America and later from China. Recent collections, made as part of the International Conifer Conservation Programme, launched in 1991 in response to the growing international concern for the plight of the world's conifers, add new diversity and scientific rigour to the historic backbone of our Gardens. Old and new provide an invaluable resource for research and conservation.

above Purple cones of beautiful silver fir *Abies forrestii* named in honour of the great early 20th century Scottish explorer George Forrest. Augmenting the historic collection at Dawyck, this specimen was grown from seed collected in Sichuan, China on the north side of Mt Luoji at 3,570m in 1990.

Gingers

Gingers are one of the Garden's major research groups where both traditional herbarium-based and more modern molecular-based techniques are used in an innovative integrated approach to research. Found in tropical forests, the ginger family, Zingiberaceae, produces spectacular flowers and their diversity makes them an interesting subject for both floristic and evolution research. RBGE produces floristic accounts of gingers for Floras of countries such as Indonesia and Thailand. Gingers, including ginger, turmeric and cardamom, are of great economic and medicinal importance.

above A spectacular display from *Curcuma zanthorrhiza*, a member of the ginger family (Zingiberaceae), found in the tropical forests of India and collected in 1999. The genus *Curcuma* includes many species of ethnobotanical interest, not least *C. longa*, turmeric, the yellow spice with anti-inflammatory properties, used in medicine and Indian cooking.

Chile

Chile is a long, narrow country of extraordinary climatic extremes, with coastline to the west, mountains to the east, desert to the north and glaciers to the south (plus a few volcanoes for good measure!). It is home to numerous well-known garden plants as well as seven threatened conifer species, which is the main reason for RBGE's original research in Chile. That has led to recent collaborative

above The Chilean lantern tree, *Crinodendron hookerianum*, an evergreen with cascades of crimson flowers, is endemic to Chile. Seen here growing outside at Logan, this specimen was grown from seed collected at 600m in the Cordillera de Los Andes, on the Universidad de Chile and RBGE Expedition to southern Chile in 1998.

Scotland

Compared with many other countries, Scotland has a relatively poor flora, although it is very rich in mosses, liverworts and lichens. But many native plants here are on the edge of their European range, which makes them both vulnerable and an interesting subject for research. Conservation is an important theme of geographic displays featuring Scottish plants that are rare or threatened in the

above Purple saxifrage *Saxifraga oppositifolia* is found on Ben Lawers, one of the highest mountains of Scotland's southern Highlands and designated a National Nature Reserve for its abundance of rare alpine plants. The arctic alpine seen here in Edinburgh can be found right around the northern latitudes of the northern hemisphere.

Geographic

Plants from across the world thrive in Scotland's surprisingly varied climate. With four Gardens, RBGE takes full advantage of regional differences in temperature and rainfall to create different geographic displays. The plants growing in the Living Collection are the result of collaborative research and conservation projects in many countries and regions of the world. Among them are Chile, Scotland, China and South West Asia.

China

Edinburgh has been a long-term centre for research on Chinese plants with globally important collections dating back to the late 19th century. Several of China's now venerated botanists visited Edinburgh in the 1930s and when China reopened in the 1980s, RBGE was one of the first institutes to be invited back. The result is continuing research collaboration on the *Flora of China* and the establishment of the Jade Dragon Field Station – a collaboration between RBGE, Kunming Institute of Botany and the Lijiang Alpine Botanic Garden.

above The handkerchief tree *Davidia involucrata*, first discovered by the French missionary Father Armand David in 1869, was collected and introduced into cultivation by Ernest Wilson in 1904. Recent Sino-British expeditions have introduced new specimens such as this tree growing in Edinburgh from wild collected seed.

South West Asia

RBGE's long history of working in South West Asia has included coordination of the production of the *Flora of Arabia* and led to the recent establishment of the Centre for Middle Eastern Plants. In a region perhaps surprisingly rich in endemic plant life, work continues with collaborative programmes for research, training, capacity building and biodiversity assessments, helping to establish botanic gardens and supporting conservation initiatives, especially in Saudi Arabia, Oman, Yemen, Iraq and Afghanistan.

above *Begonia socotrana*, a winter-flowering begonia first collected from Soqotra by Isaac Bayley Balfour in 1880, was used to breed our popular houseplant hybrids. The species, endemic to Soqotra, was recently re-collected by Tony Miller for the first time since the Balfour introduction as part of the collaborative production of the *Flora of Arabia*. This particular individual has been grown from seed collected in a north-facing limestone cleft in 1999.

Arid lands

Almost a third of Earth's land surface is arid or desert land (with annual rainfall of less than 25mm). Plants have evolved a fascinating range of shapes and forms to enable them to survive in dry climates – tiny leaves with hairy surfaces to reduce evaporation and spiny armour to deter grazing. The Garden works in many arid countries including Iraq, Afghanistan and Oman, all of which have fascinating floras. The Arid Lands House at the Edinburgh Garden features the ingenuity of plants, together with stories of equally ingenious people who live in the world's desert regions.

above   Agave sebastiana from Mexico growing in the Arid Lands House exhibits its many xerophytic adaptions allowing it to tolerate desert conditions. Spiny leaves deter herbivores, thick leaves store water and the waxy leaf surface reduces water loss.

Alpines

Alpine plants have been a special feature of RBGE for many years. The Rock Garden, established more than 100 years ago, is regarded as one of the great rock gardens of the world, enabling visitors to explore the flora of mountains from New Zealand to North America via Europe and South Africa. The Alpine House shows how plants are equipped to survive at high altitude – dwarf forms protecting against wind, snow and cold; small leaves reducing heat loss – and gives visitors a chance to see the exquisite, tiny flowers in close-up.

above   Cushioned against the cold: *Dionysia tapetodes*. The Alpine House allows a close-up view of the tiny yellow-flowered alpine, a member of the *Primula* family, displaying adaptations to mountain top conditions. Collected as seed in Iran, 1993.

Ecological

Assembling and exhibiting plants in the Living Collection is not just about creating taxonomic or geographic displays. Sometimes the Curator sets out to mimic nature by placing plants as they would be found in the wild. Plants grow well together when they are given freedom to develop in a 'natural' environment – arid, alpine, woodland and aquatic – revealing their extraordinary ability to adapt and evolve.

Woodland

An understorey of plants thrives in the dappled shade of the woodland floor. All four RBGE Gardens display a rich variety of woodland plants. In the hillside settings of Benmore and Dawyck, bulbs, ferns and herbaceous plants grow in the apparently natural environment of historic plantations. Edinburgh's urban site includes a Woodland Garden and a Copse. At Logan the Australasian woodland of tree ferns and *Eucalyptus*, interwoven with *Leptospermum* and *Grevillea* and other flowering shrubs, has an atmosphere all its own.

above Adding drama to dappled shade, giant Himalayan lilies, *Cardiocrinum giganteum*, can grow to a height of 3m. Seen here in Edinburgh's Copse but also growing at Logan, these magnificent plants thrive in rich, damp, leafy soil. Grown from seed collected on the 1996 Sichuan Expedition.

Aquatic

Around 70 per cent of the world's surface is covered by water – not just salty oceans but freshwater rivers and lakes too. Not surprisingly many plants live in water and have developed all sorts of adaptations, such as floating leaves and flexible stems, that allow them to survive. All four Gardens and the glasshouses have water features, from natural streams to formal ponds that and celebrate the great diversity of aquatic plants, not just on, or below, the surface of ponds and streams but also in bogs and water margins.

above Lotus (*Nelumbo nucifera*), the beautiful aquatic lotus, has great cultural significance in the east where it is a sacred symbol and inspiration in art and architecture – the microstructure of the self-cleaning lotus leaves inspired the invention of self-cleaning glass.

Collectors and explorers

Today's Living Collection is based mostly on wild origin plants. There are many contemporary stories to tell of botanists and horticulturists working in the field to bring back plants for study and enjoyment. But there are also great stories from the past. During the first three decades of the 20th century, George Forrest, perhaps the Garden's most iconic collector, travelled seven times to China. He died there after a heart attack in 1932. His legacy is thousands of garden plants including *Primula*, *Lilium* and *Meconopsis* as well as *Rhododendron*.

above A dwarf creeping shrub, *Rhododendron forrestii* ssp. *forrestii* bears the name of the Scottish collector George Forrest. The plant was raised from seed that he collected from damp, stony mountain pastures on the Londjre Pass, Mekong–Salween divide at 4,250m in Yunnan province, China in 1921. Between 1904 and 1932 Forrest discovered over 1,200 plant species new to science.

Orchids and cycads

Orchids and cycads illustrate some of the diversity of plant evolution. Cycads are among the most ancient land plants and were particularly abundant during the Jurassic period (about 160 million years ago). These massive, slow-growing and long-lived plants are either male or female (dioecious), and produce large distinctive cones. Orchids originated more recently, in the Cretaceous period, and have evolved highly specialised flowers that depend on a fascinating relationship with their pollinators, so specialised that if one

dies out the other almost certainly does too. Orchidaceae is the largest plant family, with about 25,000 species. In the Orchid and Cycad House these two remarkable groups of plants are displayed together for comparison.

above Telling the story of plant evolution in the glasshouses: *Cycas revoluta*, a cycad collected as seed from a cultivated plant in Japan in 1976, is one of the oldest forms of plant life.

Thematic

Plants have all kinds of stories to tell. Complex stories of evolution and simpler tales of human adventure can both be illustrated with vivid displays of plants. The Living Collection has an important part to play in assisting RBGE to explore and explain the world of plants and why human life depends on it. But spectacular displays of plants are also there for the sheer pleasure of seeing, smelling and being among them.

Floral display

Each of the four Gardens sets out to delight and amaze visitors with dazzling floral displays. Benmore, Logan and Dawyck all have their special areas of seasonal colour (the half hardy bedding at Logan, *Primula* at Benmore and drifts of blue poppies at Dawyck) but perhaps nowhere does it better than Edinburgh's Herbaceous Border – and the more recent Queen Mother's Memorial Garden is another highly popular corner of the Garden. All of these spectacular productions have just one purpose: they are purely for pleasure.

above Purely for pleasure: Edinburgh's famous Herbaceous Border is 165m long and planted to create a rainbow of summer colour – from hot reds through to cool blues and whites – with the biggest display timed for the busy months of July and August.

Exotic Logan

Plants display endless diversity in their adaptation to different habitats. From huge Victoria waterlilies and giant redwoods to diminutive orchids and alpines, the world is full of intriguing and bizarre plants. Logan has more than its fair share of nature's weirdest and most wonderful inventions – lobster claw (*Clianthus puniceus*) from New Zealand, Chatham Island forget-me-not (*Myosotidium hortensia*) from the island off New Zealand and Mount Teide bugloss (*Echium wildpretii*) from Tenerife – all bizarre and curious in their own way.

above Weird and wonderful, one of the many spectacular plants at Logan. *Fascicularia bicolor* ssp. *canaliculata* was grown from seed collected from the southern shore of the Reloncaví Estuary, Chile in 1996. In the wild this epiphytic bromeliad grows on trees. At Logan it thrives on the ground in the sheltered environment of the Walled Garden, happily placed to catch the eye.

How the Living Collection grew and grew

In the first 250 years of its existence the Royal Botanic Garden Edinburgh travelled roughly 5km across the capital city. It was a deceptively slow journey. Between 1670 and 1820 there were three moves involving at least five different garden sites as the first small plot of medicinal herbs grew to become one of the largest and most important botanical collections in the world. The Garden began during an age of enquiry and discovery and an adventurous quest for knowledge has driven each new development ever since.

left Cultivars of *Helenium* and *Lythrum* provide a bright display in the Herbaceous Border.

As the Garden moved across the city, plant collectors were travelling across the world. In the 17th century the Garden already grew plants from as far away as the Cape of Good Hope and the West Indies. By the time it arrived at Inverleith in the early 19th century a greatly increased range of plants growing under glass and outdoors reflected Britain's colonial expansion east, west and south. It was in the early 20th century that the opening up of China allowed the biggest expansion of RBGE. During the next hundred years the Living Collection grew beyond Edinburgh, as RBGE established three new botanic gardens across Scotland, each with a distinct character and microclimate.

All great gardens delight in growing a dazzling display of plants. What distinguishes botanic gardens is that this living material is meticulously documented; the gardener's skill an essential means to the end of scientific research and experiment. What sets RBGE apart from many other national botanic gardens is the global range and diversity of its world-class Living Collection.

For a small country Scotland can produce a surprisingly wide range of climatic conditions, which RBGE exploits to the full. The Regional Gardens form a botanical trio of astonishing contrasts: temperate rainforest Benmore in Argyll, sub-tropical Logan in Galloway and cold continental Dawyck in Peeblesshire. In a triangle of extremes – wettest, warmest and coldest – the longest distance is the journey between Dawyck and Logan and even that is less than 200km.

Climate and weather largely determine what survives but, as with all great collections, the legacy of plants in each Garden is the result of human choices, expertise and influential contacts. Each new era has been directed by the personalities and passions of individual Regius Keepers (a keeper appointed by royal warrant), Curators and collectors. Behind almost every plant or group of plants there are human stories. At times the history of the Living Collection reads like an adventure story following ship's surgeons and botanists exploring distant shores. Early plant catalogues reveal a fascinating network of gardeners, botanists and plant collectors connecting RBGE with the rest of the world. From their base in Edinburgh, the Regius Keepers have always been in touch with other great gardens and gardeners, sharing seeds, knowledge and an insatiable passion for plants.

opposite Across the city: a map of Edinburgh in 1837, by Robert Stevenson and Son, shows the location of former sites of the Royal Botanic Garden Edinburgh.

1 1670 Physic Garden established at Holyrood

2 1675 Move to Trinity Hospital site

3 1684 Extended to Trinity College Kirkyard

4 1763 Move to Leith Walk

5 1820 Move to Inverleith

above Growing across Scotland: the location of the four Gardens of RBGE.

1 Edinburgh

2 Dawyck

3 Benmore

4 Logan

Together they exploit to the full the diverse climate, topography and soils of Scotland, allowing the cultivation of one of the largest collections of species-based plants in the world.

above right Global reach: A map showing just some of the countries in which RBGE works. The Garden has working partnerships with over 40 countries and links with a further 40.

Afghanistan	Chile	India	Nepal	Congo	Syria
Argentina	China	Indonesia	Netherlands	Russia	Taiwan
Armenia	Colombia	Iran	New Caledonia	Saudi Arabia	Tanzania
Australia	Costa Rica	Iraq	New Guinea	Scotland	Thailand
Austria	Czech Republic	Ireland	New Zealand	Serbia	Turkey
Azerbaijan	Ecuador	Israel	Oman	Singapore	Turkmenistan
Bahrain	Egypt	Italy	Pakistan	Slovakia	UAE
Belgium	England	Japan	Panama	Solomon Islands	Ukraine
Belize	France	Jordan	Peru	Somalia	USA
Bhutan	Gabon	Kazakhstan	Philippines	South Africa	Vietnam
Bolivia	Georgia	Kuwait	Poland	South Korea	Wales
Bosnia	Germany	Lao PDR	Portugal	Spain	Yemen
Brazil	Greenland	Lebanon	Qatar	Sweden	
Cambodia	Hungary	Malaysia	Republic of	Switzerland	

Accident and natural disaster also played their part in shaping the Living Collection. Changing plants in the Gardens reflected changing social, political and commercial interests as well as changing family fortunes. Benmore – the only Regional Garden that RBGE deliberately set out to acquire – was bequeathed to the nation by the Younger family as the golden era of great estates came to an end between the two World Wars. RBGE fell heir to the McDouall's southern hemisphere collection at Logan for similar reasons in the late 1960s and the hurricane that roared across central Scotland in 1968 helped to speed the Balfour family's gift of Dawyck.

A garden is a dynamic environment. The Living Collection has constantly responded to changes in the wider world: from the Enlightenment of the 18th century through the 19th century Industrial Revolution to a new awakening in the late 20th century. With new understanding that loss of bio-diversity is as great a threat to humanity as climate change, botanists and gardeners are once again at the forefront of discovery and research. Today's display and diversity of plants reflect a new era of exploration and international collaboration (RBGE currently has active partnerships with 44 countries and links with a further 40). Recent naturalistic geographic plantings in each of RBGE's four Gardens are the result of new expeditions and a strategically focused plant collection policy which concentrates resources where they are needed most, setting priorities for new plants coming into the collection. More than ever before, the display of the Living Collection sets out to engage, excite and inform the public.

Together, the four Gardens of RBGE grow an invaluable resource for research, conservation, education and enjoyment. The Living Collection holds in safe-keeping almost 35,000 different plant collections – almost 15,000 different species, perhaps 5 per cent of the world's known plant species – many of them threatened in their natural environment. Today's rapidly changing world is very different from the early Enlightenment of RBGE's founding fathers, Robert Sibbald and Andrew Balfour. Yet the fundamental purpose of the Living Collection – to collect, grow and study plants for human health and well-being – is essentially unchanged since two doctors sowed the first seeds on a small square of land in Holyrood Abbey not very far from today's Scottish Parliament. After 340 years the story is still unfolding.

Key events and influences in 340 years of the Living Collection.

1670

Robert Sibbald and Andrew Balfour lease a small 0.03ha plot near Holyrood Abbey, stocked with plants from their own gardens, to instruct students in botany, train apothecaries and provide the foundation of a Scottish Pharmacopoeia.

1675

The Physic Garden is established on a 2ha site in Trinity Hospital grounds near the marshy Nor'Loch (now Princes Street Gardens).

1676

James Sutherland is hired to care for the growing collection including tender plants under glass frames.

1683

Sutherland publishes the Garden's first catalogue, *Hortus Medicus Edinburgensis*, with a list of 2,000 plants including most of the herbs used in medicine as well as Scottish native plants and garden varieties.

1689

Most of the Trinity Garden plants are destroyed in a flood when a dam is breached during a siege of Edinburgh Castle in the Glorious Revolution (which results in James II being replaced by William and Mary).

1699

Sutherland is named King's Botanist in a Royal Warrant from William III. [During the early 1700s Sutherland takes charge of three botanic or physic gardens in the city: one around Holyrood, known as the King's Garden, one around Trinity Hospital, known as the Town's Garden, and one beside the College, the College Garden.]

1753

In Sweden Carl Linnaeus publishes *Species Plantarum* using Latin names in two parts: first the genus, followed by the species.

1763

John Hope unifies the Garden on a 2ha site in Leith Walk, outside the city boundary, with glasshouses, pond and arboretum. In the same year Hope forms the first British syndicate for importing foreign seeds and plants.

1775

The Garden Catalogue reveals the growing collection and the influence of Hope's Society for the Importation of Foreign Seeds and Plants, linking Edinburgh with a global network from Far East to Far West and with almost 70 per cent of new plants from North America.

1782–1795

The Scottish explorer Archibald Menzies introduces many North American plants to Britain.

1817

Rapidly outgrowing the Leith Walk site, Regius Keeper Daniel Rutherford considers a move to Holyrood with Salisbury Crags as a natural rock garden for the collection of Cape heaths.

1820–1823

The Garden moves 2.4km north to a 5.9ha site at Inverleith, bringing mature trees and plants with William McNab's specially invented horse-drawn transplanting machine.

1826–1835

David Douglas explores north-west America, introducing conifers and shrubs that transform British gardens and woodlands.

1832–1836

Charles Darwin's voyage on the *Beagle*.

1834

The opening of the octagonal Tropical Palm House, 15m high and the largest in Britain; steam-heating produces vigorous growth in Malayan palms which will literally go through the roof.

1858

The completion of the Temperate Palm House gives breathing space in the now overcrowded older Palm House.

1864

The Garden acquires 4ha from the Royal Caledonian Horticultural Society, gaining space for James McNab's Rock Garden in the early 1870s.

1877

The purchase of Inverleith House and grounds gains 12ha for the Arboretum.

1899–1910

Ernest 'Chinese' Wilson introduces rhododendrons, herbaceous plants and trees from his travels to China, Japan, Korea, Taiwan and Australasia.

1903

Sir Isaac Bayley Balfour, Regius Keeper, hires George Forrest as apprentice in the herbarium and recommends him for A.K. Bulley's expedition to western China.

1904–1930

George Forrest explores Yunnan in south-west China, Tibet and Burma, introducing hundreds of plants new to science and establishing RBGE's internationally renowned rhododendron collection.

1929

Younger Botanic Garden (now Benmore), the first RBGE 'outstation', provides 44ha of Argyll mountainside for a huge influx of Chinese plants arriving in Edinburgh.

1911–1956

Frank Kingdon Ward explores China, Burma, Tibet and Assam, introducing many new plants including *Meconopsis betonicifolia.*

1969

Logan Botanic Garden, 12ha in Wigtownshire, becomes part of RBGE.

1970s

The Living Collection expands with a new era of RBGE botanical expeditions. The RBGE *Collection Policy* now emphasises the importance of wild origin plants over garden exchanges.

1979

Dawyck Botanic Garden: RBGE acquires 25ha of hillside arboretum in Peeblesshire.

1980s

New Sino-British expeditions bring a massive new influx of plants from China to RBGE.

1991

RBGE launches the International Conifer Conservation Programme in a decade of increasing conservation partnerships across the world.

1992

The Earth Summit in Rio on environment and sustainable development sets the agenda for United Nations conventions on climate change and biological diversity.

2001

The foundation stone is laid for Lijiang Botanic Garden in China, run between RBGE, the Kunming Institute of Botany and the Yunnan Academy of Agricultural Science.

2002

The United Nations Convention on Biological Diversity launches the Global Strategy for Plant Conservation. Of the seventeen targets eight are highly relevant to the work of botanic gardens.

2009

Benmore's Victorian Fernery is restored and opened to the public.

2009

A new visitor centre incorporating a shop, a restaurant and exhibition space is opened at Dawyck and, a year later, Dawyck is awarded 5 star tourist attraction status, the first garden to be so awarded in Scotland.

2010

The John Hope Gateway and Biodiversity Garden are opened by the Queen at the West Gate of RBGE. Together the four Gardens of RBGE grow almost 35,000 different plant collections (around 15,000 species from all over the world, or about 5 per cent of all known plant species).

2010

The Tenth Conference of the Parties to the Convention on Biological Diversity calls for international government cooperation to halt and reverse the loss of biological diversity.

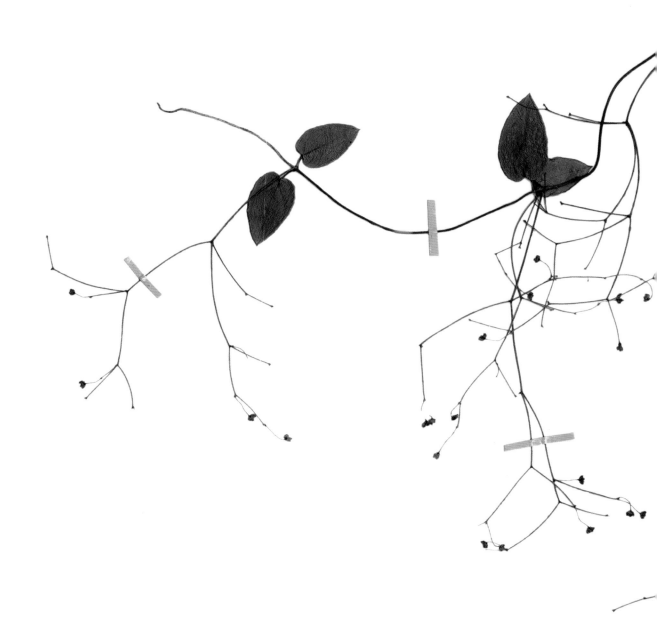

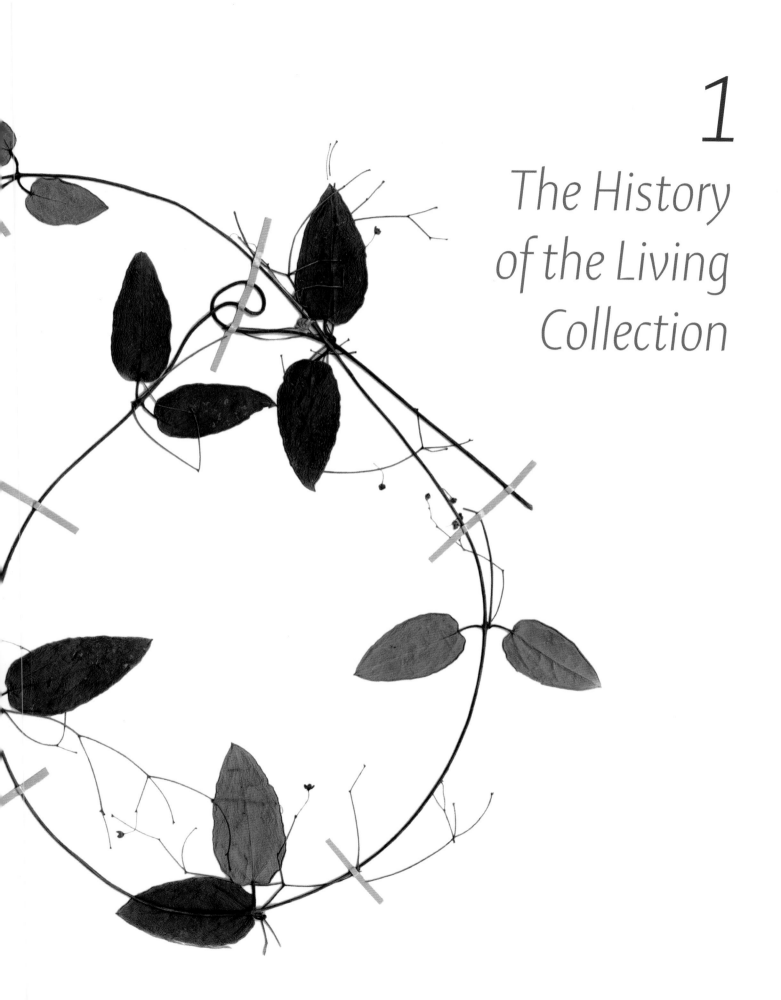

The History of the Living Collection

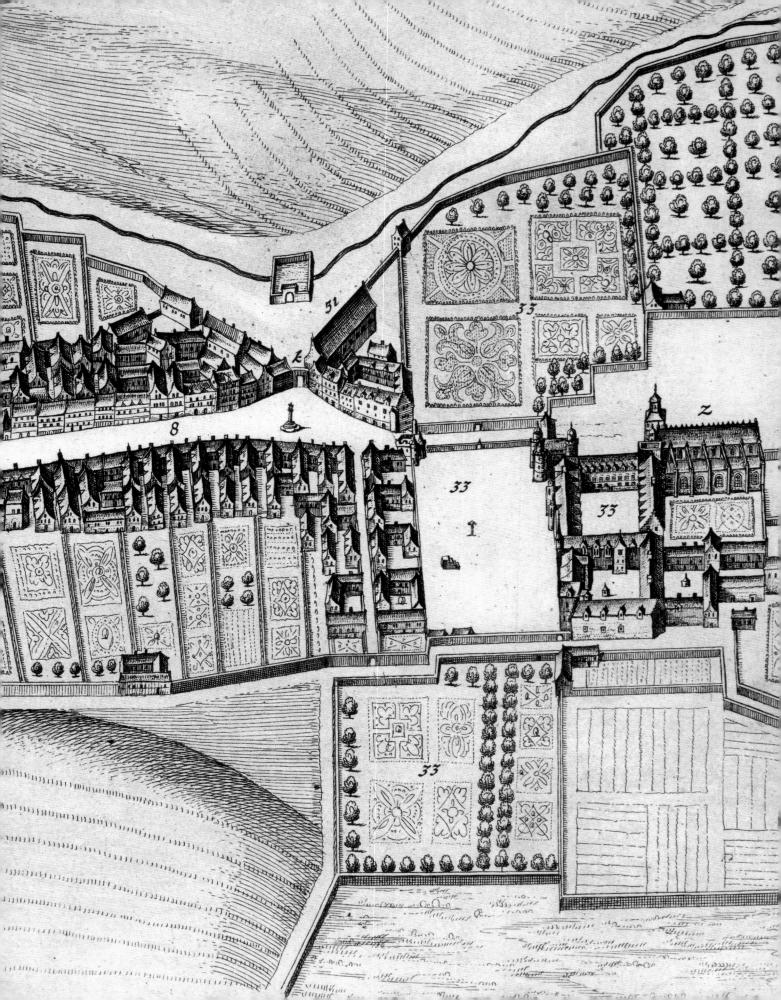

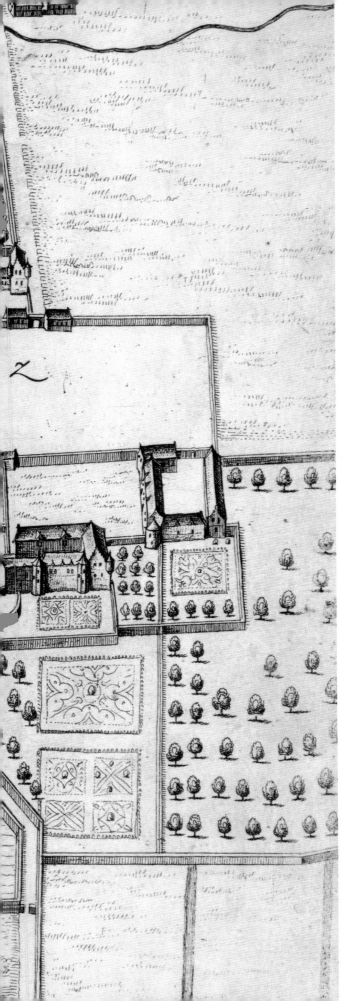

Scotland's first botanic garden

The story begins with medicine. In 1670 two doctors started a collection of medical plants on a small plot of land on the edge of Holyrood Park. The aim of Robert Sibbald and Andrew Balfour was to grow living material for investigation, experiment and teaching. It was a bold step by two physicians who were dissatisfied with the state of medicine in Scotland.

left Although the exact site is not known, an early engraving by James Gordon in 1647 shows a number of garden plots around the Royal Palace of Holyrood House, one of which may well have become the first Physic Garden in 1670.

In the second half of the 17th century Scotland was a poor country. Yet poverty combined with a rigorous education system to produce ambitious and adventurous scholars who travelled across Europe. Sibbald and Balfour were part of an international network sharing the knowledge and ideas of the early Enlightenment. Impressed by the quality of botanical education in European universities, the doctors returned from their travels determined to raise standards at home. They brought with them not only seeds of plants unknown in Scotland but also the seed of a garden that would grow to become one of the largest and most important living collections in the world.

The small square plot in St Anne's Yard by Holyrood Abbey helped to establish Edinburgh's reputation for medical excellence. But it was also the foundation of Scotland's first botanic garden and only the second in Britain (only Oxford, founded in 1621, is older). On a piece of land 12m square, barely bigger than a tennis court, Sibbald and Balfour began

A Renaissance spirit of enquiry

The first modern botanic garden for which records survive was established at the University of Pisa in 1543 by Luca Ghini (1490–1556). This was followed by the gardens of Padua in 1544 and Florence a year later. All three gardens were established with the help of the wealthy Medici family of merchants and bankers based in Florence.

In the history of the modern botanical garden, this period is known as the 'European Period' and further examples of gardens from this era include Bologna (1547), Zurich (1560), Leiden (1577), Heidelberg (1593), Montpellier (1598), Oxford (1621), Edinburgh (1670), Berlin (1679) and Amsterdam (1682). Paris was founded as a Royal Garden in 1597, but only became a botanical collection as the Jardin du Roi in 1626 (later called the Jardin du Roi des Plantes). All these gardens reflected the Renaissance spirit of enquiry, the need to gather knowledge for knowledge's sake.

The functions of these medicinal or physic gardens were initially restricted to the growing of medicinal plants for teaching and use. However, plants with little-known properties, medicinal values or other attributes were introduced into these gardens over time, and it became a legitimate interest for such plants to be investigated. Before long it became normal practice to grow increasing numbers of plant species in botanic gardens for scientific study. From these original medicinal foundations, many of today's major European botanic gardens have arisen.

right Engraving of Padua Botanic Garden made in 1654. The 16th century design, a square within a circle, is still in existence today and Padua is the world's oldest intact botanic garden remaining in its original location.

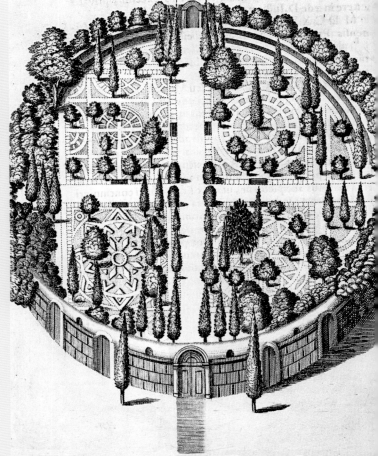

VIRIDARIVM GYMNASII PATAVINI MEDICVM

Founding fathers

The first seeds of the Living Collection were sown by two remarkable doctors, Andrew Balfour and Robert Sibbald.

Balfour (1630–1694), the first doctor in Scotland to dissect a human body, was a distinguished academic and collector of scientific instruments. His studies took him to the Jardin du Roi in Paris during the 1650s. After returning to St Andrews in 1667 he moved to Edinburgh where he stocked his own small garden with rare plants sent by his foreign correspondents. Balfour was largely responsible for founding the Royal College of Physicians of Edinburgh in 1681 and became its third President in 1684. He was knighted in 1682.

Robert Sibbald (1641–1722) was born and educated in Edinburgh and studied medicine in Leiden, Paris and Angers. Returning to Scotland, he contrasted the poor state of medicine and lack of pharmacopoeia (a reference work for the raw materials of medicinal compounds) with higher European standards. He too created a small private garden in which to grow medicinal herbs. Charles II made Sibbald King's Physician, Geographer Royal and Natural Historian. Sibbald was knighted in 1682 and in 1684 he was elected the second President of the Royal College of Physicians of Edinburgh.

Balfour and Sibbald met when they returned to Scotland and agreed that Edinburgh should have the kind of physic garden they had seen in Europe. In 1670, they leased a small plot of land near Holyrood Abbey and began to plant.

right Robbert Sibbald. This portrait by John Alexander is in the Royal College of Physicians of Edinburgh.

opposite page, top
Holyrood House Palace today. At the bottom of the Royal Mile in the heart of Edinburgh's Old Town, this was the location of the first Physic Garden, founded on a plot called St Anne's Yard.

what was to become a world-famous collection with a mixture of mostly medicinal plants of between 800 and 900 species and horticultural varieties. Some of them were transplanted from their own private gardens, some were collected wild from the Scottish countryside, many more came from their friend Patrick Murray, the laird of Livingston in West Lothian, a landowner who shared their passion for natural history, plants and travel.

The purpose of the garden was to teach students botany and train apothecaries in plant-based *materia medica* (the ingredients for making medicines). But no gardener can resist the temptation to try growing exciting new specimens. Once James Sutherland was appointed to care for the plot the collection grew rapidly, again with the help of international connections. The man who would become the first King's Botanist corresponded with botanists across Europe to the East and West Indies, obtaining seeds from many of them.

Not much is known about Sutherland's background beyond the fact that he was a skilled gardener and keen collector of coins – Sibbald's memoirs describe him as a youth who had "attained great knowledge of the plants and of medals" – and clearly he had a passion for plants. Sutherland collected and grew just about everything that was available at the time, whether it was medicinal or not. Under his care the Garden grew so well the Holyrood Abbey site soon became too small, and in 1675 the Living Collection began its first move across the city.

The Physic Garden at Trinity Hospital

Andrew Balfour, backed by most of the medical profession in Edinburgh, successfully petitioned the Town Council to lease Sutherland a bigger plot. The new site, known as the Physic Garden, was at Trinity Hospital at the mouth of the Nor'Loch, a marshy area to the north-east of Edinburgh Castle (now hidden beneath Waverley Station). The Physic Garden was divided into six rectangles, three on each side of a canal draining the marsh, including borders for garden flowers, trees and shrubs as well as beds for medicinal plants. There were bell-jars, frames and walls to give protection to more delicate plants.

As Sibbald recorded, plants were imported "from all places … considerable packquets of seeds and plants were yeerly sent hither from abroad". By the time Sutherland published his *Hortus Medicus Edinburgensis* in 1683, the Physic Garden collection had grown to 2,000 plants coming from as far away as the Cape of Good Hope and the West Indies. The subtitle of the first publication of its kind in Scotland was *A Catalogue of the Plants in the Physical Garden at Edinburgh; containing Their most proper Latin and English names; With an English Alphabetical Index.* Sutherland was evidently keen to make the book accessible to a wide range of students and his preface

reveals a moving personal belief in the value and significance of the medicinal plants in the collection. He notes that many of the plants "that were wanting here and therefore yearly brought from abroad because of their usefulness in Physic, may now by Industry and Culture be had in plenty at home. Before the establishment of the garden the Apothecaries' Apprentices could never be competently instructed, as they should be … Now they could learn more in the space of one summer than it was formerly possible for them to do in an Age."

He could not know that disaster was looming. The Garden was flourishing against a backdrop of religious wars between opposing factions within the British monarchy. In 1684 the Physic Garden grew into Trinity College Kirkyard but in 1689 the dam between the Nor'Loch and its drainage

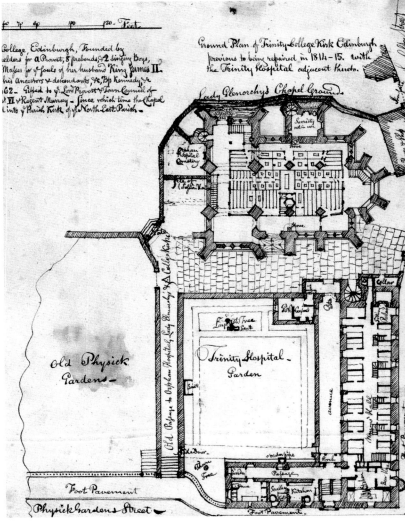

above In 1684 the Physic Garden became too small and expanded into Trinity College Kirkyard, adjacent to Trinity Chapel depicted here.

right Plan showing Trinity College Kirk (at the top of this plan) adjacent to Trinity Hospital. This plan, drawn in 1814–1815, shows a plot marked "Old Physick Gardens".

NEAR THIS SPOT FROM 1675 TO 1763
WAS THE EDINBURGH PHYSIC GARDEN,
ORIGINALLY FOUNDED AT HOLYROOD IN 1670
BY
SIR ROBERT SIBBALD AND SIR ANDREW BALFOUR, TWO OF
THE FOUNDERS OF THE ROYAL COLLEGE OF PHYSICIANS OF EDINBURGH.
THE GARDEN, UNDER THE CONTROL OF JAMES SUTHERLAND
THE FIRST REGIUS PROFESSOR OF BOTANY IN THE UNIVERSITY,
WAS THE DIRECT PREDECESSOR OF THE PRESENT
ROYAL BOTANIC GARDEN.

THIS PLAQUE WAS ERECTED IN 1978
BY
THE ROYAL COLLEGE OF PHYSICIANS OF EDINBURGH,
THE UNIVERSITY OF EDINBURGH AND THE
ROYAL BOTANIC GARDEN.

channel burst when Edinburgh Castle was besieged. The almost bloodless Glorious Revolution overthrew the Catholic Jacobites and brought the Protestant William and Mary to the throne. However the siege of Jacobite supporters in the Castle caused a flood which left a thick layer of mud over the Garden that took a season to clear and destroyed virtually all the plants. Perhaps that was a turning point for Sutherland. For a while his fortunes continued to grow as he rebuilt the collection. In 1695 the Town Council created a Chair of Botany for him, a role that came with responsibility for planting the Town's College Garden and teaching at the College. The same year he also became supervisor of the King's Garden at Holyrood. Four years later, on 12th January 1699, Sutherland was named King's Botanist by William III, gaining the Royal Warrant which is the origin of RBGE's status as a royal garden.

In the early years of the 18th century there was a confusion of botanic garden sites in Edinburgh – the Royal Garden at Holyrood, the Town's Garden at Trinity Hospital and the College Garden beside the Town College – and all of them were under the care of James Sutherland. Even so, he seemed to

lose enthusiasm for plants and gardens and turned to his other passion, collecting coins. In 1706, at the age of 67, after repeated complaints from Town Council and University, Sutherland resigned from his University post as Professor of Botany. However, he retained his title as Regius Keeper and King's Botanist.

Despite his unceremonious departure, James Sutherland left a legacy that has stood the test of time. His *Hortus Medicus Edinburgensis* offers a fascinating insight into 17th century horticulture in Scotland. About a fifth of the plants were 'officinal' (used in medicine), while the rest were ornamental or 'botanical', growing wild in Scotland. Sutherland's list of ornamental garden plants included almost everything to be found growing in Scottish estate gardens at the time in addition to quite a few not previously recorded. There were ten cultivars of spring crocus (*Crocus vernus*), five of autumn crocus (*Colchicum autumnale*), five of crown imperial (*Fritillaria imperialis*) as well as numerous cultivars of carnation, daffodil, lily, scilla, tulip, rose and more. Sutherland's sources reveal a network of friends and keen collectors at home and abroad who

above Plaque in Waverley railway station commemorating the site of the Physic Garden when it was at Trinty College.

Herbal healing

James Sutherland's *Hortus Medicus Edinburgensis*, published in 1683, lists 383 species that were primarily 'officinal' or medicinal species. John Gerard had published *The Herbal or General History of Plants* in 1597 and in 1633 Thomas Johnson revised and enlarged it. Each entry in the *Herbal* includes an engraving of the plant, a description and a list of its 'virtues' or uses. As Sutherland's *Hortus* appeared only 50 years after the publication of the revised *Herbal* it is not unreasonable to suppose that medicinal uses of the plants remained the same.

These illustrations and citations from the *Herbal* are of plants known to have been grown in Sutherland's Physic Garden at Trinity Hospital as they are listed in his *Hortus*.

Scots Pine (*Pinus sylvestris*)
(below right)

"Being stamped and boyled in vinegar, they assuage the paine of the teeth, if they be washed with this decoction hot: the same be also good for those that have bad livers, being drunk with water or mead."

Walnut (*Juglans regia*) (left)

"With onions, salt, and honey, they are good against the biting of a mad dog or man, if they be laid upon the wound."

Elderberry (*Sambucus nigra*)
(top right)

"The leaves and tender crops of common Elder taken in some broth or pottage open the belly, purging both slimie flegem and choloricke humours: the middle barke is of the same nature, but stronger, and purgeth the said humours more violently."

Bay (*Laurus nobilis*) (top middle)

"Bay Berries taken in wine are good against the bitings and stingings of any venomous beast, and against all venome and poison … the juice pressed out hereof is a remedy for paine of the ears, and deafness, if it be dropped on with old wine and oile of roses."

Thyme (*Thymus* sp.) (below middle)

"Wilde thyme boyled in wine and drunke, is good against the wambling and grippings of the bellie, ruptures, convulsions, and inflammations of the liver."

Nux Iuglans.
The Walnut tree.

Laurus.
The Bay tree.

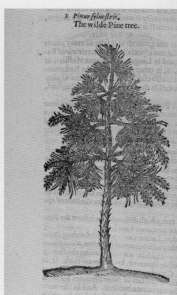

1 Sambucus.
The common Elder tree.

‡ 7 *Serpillum citratum.*
Limon Time.

1 Pinus sylvestris.
The wilde Pine tree.

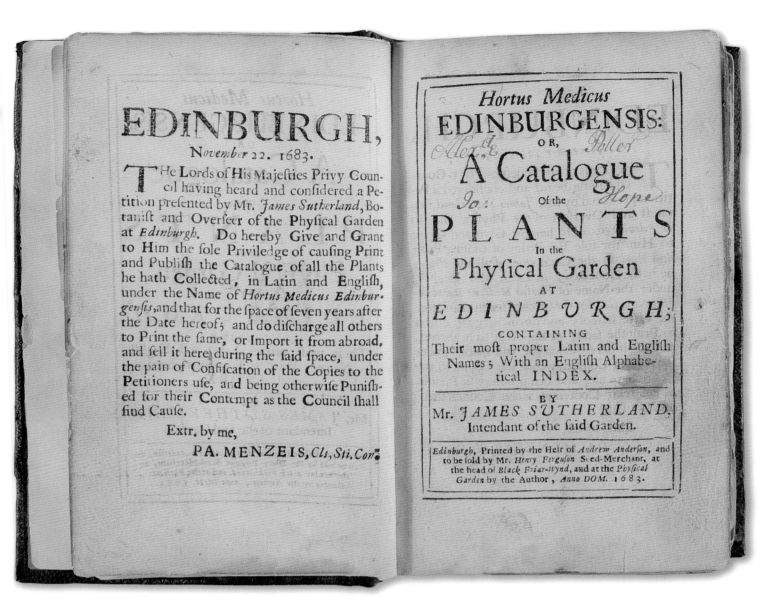

corresponded regularly and exchanged plants for their private collections (some of these letters survive and are now held in the British Library in London). He also collected plants on his own travels around Scotland but he had another important source: Balfour and Sibbald instructed their medical students to collect plants and seed from abroad and send them to Sutherland. These newly trained medics often began their careers as ship's doctors and found themselves in distant lands. This may account for many of the rather obscure plants from around the world that appear in Sutherland's catalogue.

James Sutherland died in 1719 at the age of 80. A succession of different Keepers, Curators and collectors were to make their individual contribution to the shape, diversity and colour of the Living Collection over the next 300 years. But the first Regius Keeper is rightly remembered as the man who laid the foundations of today's Garden with his own high standards in cultivation and education and an insatiable curiosity about plants.

The era of John Hope and the move to Leith Walk

The appointment of John Hope in 1761 marked a major development in the Living Collection. Hope, both a product and a pioneer of the Enlightenment, corresponded with scientists and plant collectors at home and abroad and these influences were to show not only in the plants growing in the Garden but also in the way they were ordered and classified. The dedicated doctor of medicine was also a diligent and enthusiastic botanist. During his era there was a significant shift in plant research from medicine to botany.

left A perspective view of Leith Walk Garden in 1771, eight years after moving from the Trinity College site. The watercolour shows the Gardener's Cottage on Leith Walk, with pond and glasshouse to the left and, above the Cottage, the 'Schola Botanica' growing medicinal plants.

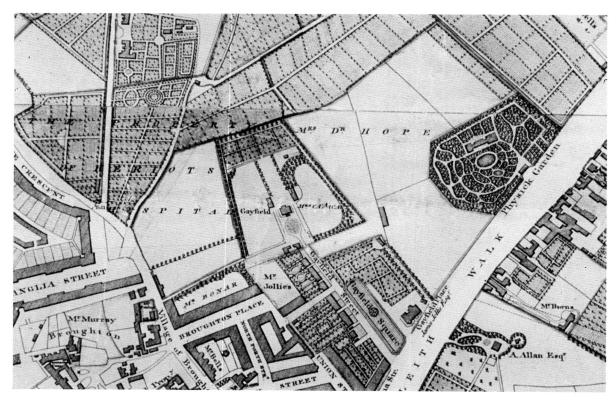

right Map of Edinburgh in 1804 showing the location of the Leith Walk Garden. Very little of this site remains except for some land within Hopetoun Crescent Garden (marked by an interpretative plaque) and the Gardener's Cottage which was demolished in 2008 after detailed archaeological investigation for possible reconstruction within the Garden at Inverleith.

From his earliest years John Hope had contact with the world of plants. He was the son of Robert Hope, a surgeon-apothecary who would have had a professional interest in plants and herbs as well as access to the two Physic Gardens in the city. Hope followed his father into medicine and attended Edinburgh University, where his enthusiasm for the science of botany was fired by the teaching of Charles Alston (his predecessor as Regius Keeper) and further developed by studying in Paris at the Jardin du Roi.

When Hope became Regius Keeper in 1761 he set out to combine the Physic Garden at Trinity Hospital and the Royal Garden at Holyrood into one, much larger, botanic

A man of compassion

John Hope (1725–1786), the son of surgeon-apothecary Robert Hope, took an early interest in botany but showed equal aptitude for medicine. In 1761 he became Regius Keeper, King's Botanist for Scotland and Superintendent of the Royal Botanic Garden in Edinburgh and was also appointed joint Professor of Botany and *Materia Medica* (the medicinal properties of substances, in this case plants) in the Faculty of Medicine in Edinburgh University.

Despite his passion for plants Hope maintained an active involvement in medicine – as President of the Royal College of Physicians he campaigned to improve health and cleanliness in the city and he was widely respected for his care and humanity towards his patients throughout his life.

He taught *Materia Medica* in the winter and botany in the summer. He was also very interested in plant physiology and conducted experiments, for example in studying the response of plants to gravity and to light, and the movement of sap and seed ripening.

left An etching of John Hope talking to a gardener (probably Malcolm McCoig, the Principal Gardener). Etching by John Kay, 1786.

garden. The Nor'Loch tended to flood the Physic Garden with sewage and was vulnerable to grazing animals. Moreover, the area was about to become a building site for the construction of the grand new North Bridge which would connect the old and new towns of Edinburgh. It was clearly time to move. With approval and financial backing from the Treasury, in 1763 John Hope transported the Living Collection to open, unpolluted ground on Leith Walk about 2km to the north-east of the city.

The 2ha site was laid out in a very different way from the earlier Gardens. During the tenure of Charles Alston the focus of plant research had been on medicine, but now Hope firmly shifted the emphasis to botanical science. His planting scheme still made ample provision for the cultivation and display of medicinal trees, shrubs and herbaceous plants but there was scope for much more variety and experiment. His glasshouses were among the earliest to be built in Scotland. There was room for forest trees and a pond for aquatics. The demonstration beds of medicinal plants occupied less than one third of the total area. Hope drew on his experience at the Jardin du Roi and even borrowed their name, *Schola Botanica*. But there was one important difference. Hope arranged species according to the Sexual System of Carl Linnaeus and also used the Linnean binomial names of plants; first the genus, then the species (a practice not fully adopted in the Paris garden until 1774). This was radical thinking at the time – the Swedish naturalist had introduced the system in 1753 – and it is partly through Hope's teaching that Linnaeus' ideas gained a firm hold in Britain.

below The Monument in the Edinburgh Garden honouring Linnaeus, commissioned and paid for by John Hope.

The naming of plants

Carl Linnaeus (1707–1778), often referred to as the 'father of modern taxonomy', was born in Småland in southern Sweden. He studied medicine and botany at Uppsala University from 1728 and for the next ten years his travels and studies brought him in contact with leading botanists, physicians and natural scientists of the day. Following a 2,000km trip across Lapland in 1732 – he returned with 100 plants new to science – he studied medicine in the Netherlands and travelled to London and Oxford before returning to Sweden in 1738.

He is best known for a number of highly significant publications in which he listed and classified plants (and animals) using a hierarchical system of classification based on a plant's floral structure. This binomial system used just two names – the genus and the species – rather than the previous form of classification which used a string of words (the 'polynomial system'). While Linnaeus did not invent the binomial system, he was the first to use it consistently. The first edition of *Systema Naturae*, published in 1735, listed over 4,000 species of animal and 7,000 species of plant while *Species Plantarum*, first published in 1753, is now regarded as the primary starting point of plant nomenclature as it is used today.

A worldwide web

In 1763–1764 Hope also initiated what was probably the first syndicate in Britain for importing foreign plants, the Society for the Importation of Foreign Seeds and Plants. This scheme grew from North American contacts (intriguingly, it involved correspondence with Benjamin Franklin). Society members paid two guineas a year for a share of all seeds or plants purchased from collectors across North America. It was clearly part of Hope's plan for improving the Edinburgh Garden but it also reflected a wider aim of introducing exotic plants of economic value, especially for use in medicine or dyeing. The Society was dissolved ten years later but it established Hope's almost worldwide network of contacts. He gained contacts closer to home in 1766 when he toured British gardens, nurseries, plant collections and libraries, visiting Cambridge, Chelsea, Kew and Oxford among others to learn more about garden management – and to find new plants for the Edinburgh Collection.

Garden catalogues of 1775 and 1778 give a fascinating insight into the greatly expanded range of plants being cultivated since Sutherland's 1683 catalogue – in alphabetical order from *Acer* (maples) of North America, Europe and western Asia to *Zanthoxylon* (the prickly yellow wood or yellow Hercules) of Jamaica. In the 1683 catalogue there were just three species of maple but by 1775 there were

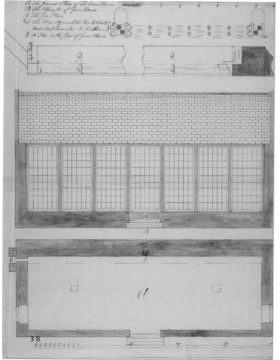

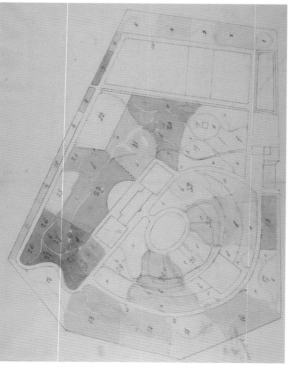

above Perspective view of the Leith Walk Garden, 1771.

far left Plan and elevation of the central glasshouse.

left Layout of the beds at the Leith Walk Garden, c.1780.

right Plan of Leith Walk Garden, dated 6 September 1777. To the right is the Gardener's Cottage where Hope gave lectures in the upstairs room. Also clearly visible is the rectangular 'Schola Botanica' for medicinal plants, the pond, glasshouse and serpentine paths and borders. But, as H.J. Noltie states in his book *John Hope (1725–1786)*, the Garden also served as a site of commemoration with different garden areas honouring botanists such as Sibbald, Ray, Tournefort and Caesalpinus, and gardeners such as Phillip Miller and Peter Collinson.

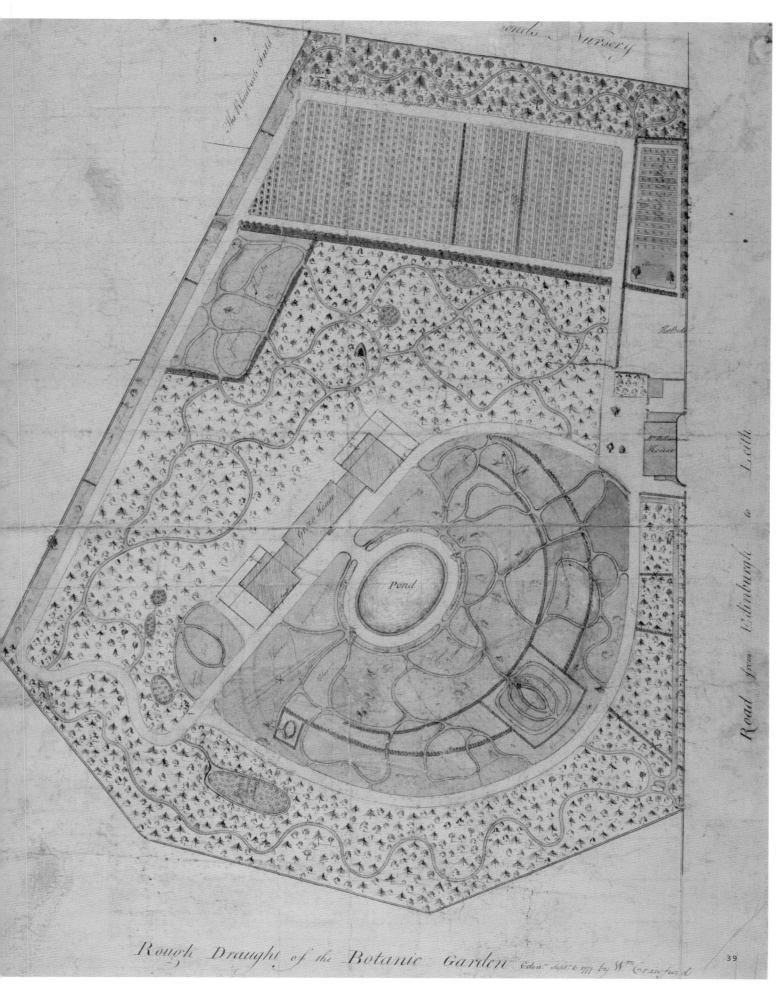

Rough Draught of the Botanic Garden Edinr Septr 6 1777 by Wm Crawford

39

a further six species – from eastern North America, southern Europe and western Africa. Individual species appearing for the first time included *Liquidambar styraciflua* (sweet gum) and *Tilia americana* (American lime). Almost 70 per cent of plants new to the 1775 catalogue were from northern America with a few from Europe and some from the UK; many of them were introduced thanks to the Society for the Importation of Foreign Seeds and Plants.

In the glasshouses an extraordinary range of tender plants were cultivated. Britain's colonial and trading interests of the time can be traced in the handwritten lists from 1774 to 1776 – about 230 species of 'Tender Exotic Plants' from South Africa, the Americas, South-East Asia, the West and East Indies, China, Japan, India, Australia, New Zealand and the Mediterranean.

At the same time Hope ensured that there was a comprehensive collection of medicinal plants for teaching and

right In memory of John Williamson, Principal Gardener with John Hope for 20 years, who died in 1780. The tablet was erected at the Leith Walk Garden above an arch close to the Gardener's Cottage and is now mounted on the wall near the Lecture Theatre.

treatment. He led some of the first experiments in the response of plants to gravity and light, the movement of sap and seed ripening. He also took a personal interest in trying to bring two frequently used foreign medicinal plants into cultivation – rhubarb, used for digestive problems, and asafoetida (*Ferula persica*), used as an anti-spasmodic. Trials of 'Turkey' rhubarb (*Rheum palmatum*) eventually proved that home-grown plants (previously considered inferior) were as effective as expensive material imported from central Asia. This explains Hope's 3,000 rhubarb plants growing just outside the Leith Walk Garden.

In all of this horticultural and botanical activity, Hope was supported by John Williamson, the Principal Gardener for 25 years. Williamson had worked at the Holyrood and Trinity gardens and his ability, industry and loyalty were an essential driving force in the layout, construction and development of the new Garden. Not only was he in charge of all the work in the Garden but he also carried out what would now be the duties of a laboratory technician, producing specimens for lectures and class exercises, carrying out and recording experiments, and preparing and storing museum exhibits. Hope valued Williamson highly and was greatly affected by his unfortunate murder in 1780, so much so that he placed in the Garden a stone tablet in his memory (still preserved in today's Garden).

Hope died in 1786, leaving a considerable legacy to the Garden (honoured today in the new Gateway building bearing his name at the West Gate in Inverleith) and to Scottish botany in general. From a manuscript of Hope's lectures in the RBGE Library it is clear that his interests were increasingly devoted to plant physiology and systematic botany. There is another monument in the Garden – his admiration for Linnaeus was so great that he erected a memorial in his honour and, characteristically, paid for it out of his own pocket.

A teacher's legacy

Of Hope's many students three stand out in botanical history: William Roxburgh, Archibald Menzies and James Edward Smith.

William Roxburgh (left) became Superintendent of the Calcutta Botanic Garden and the first botanist to attempt to draw up a systematic account of the plants of India. *Flora Indica* was to become the only book of its kind for many years until the publication of Sir Joseph Hooker's monumental *Flora of British India*.

Archibald Menzies (middle) was to introduce many North American plants to Britain during his time in the Halifax Station, Nova Scotia from 1784 to 1786. Later, from Captain George Vancouver's voyage around the world (1791–1795), he introduced the monkey puzzle tree (*Araucaria araucana*) into Britain.

James Edward Smith (right) was inspired by Hope when in 1781 he chose to study medicine in Edinburgh. Some years later he purchased Linnaeus' herbarium, papers and books and founded the Linnean Society of London. He went on to produce 36 volumes of *English Botany* (1790–1814) with illustrations by James Sowerby.

The big move to Inverleith

"The procession through the suburbs for many days ... of men and horses and waving boughs ... presented a spectacle that was at once novel and imposing".

Sir Henry Steuart in *The Planter's Guide*, 1828

left Designed for heavy lifting: although this picture was taken in the 1890s, it shows a tree transplanting machine very similar to the one used 70 years earlier when the Garden moved from Leith Walk to Inverleith. The formidable task took three years to complete and transported trees up to 11m (the tallest was a mannah ash, *Fraxinus ornus*, measuring 11.4m).

By the early 19th century the Garden was once more outgrowing its site. After Hope's death in 1786, Daniel Rutherford became Regius Keeper. He was a doctor with a special interest in chemistry, in particular the way plants reacted to the atmosphere. Rutherford appointed a succession of able Principal Gardeners (many of them keen field botanists who added to the Living Collection through plant collecting expeditions around Scotland). The most notable was the sixth appointment, William McNab, whose work was to shape the Living Collection for almost 30 years. In 1810 McNab arrived in the Garden after serving his apprenticeship at Kew to become the most energetic and longest-serving Principal Gardener (or Curator) in the history of RBGE and one of Scotland's greatest and most innovative gardeners. He introduced a wealth of new tropical and southern hemisphere plants including his personal collection of Cape heaths. During his time the Living Collection flourished as never before. In 1812 a report listed over 4,000 species with around 1,000 genera.

Rutherford began the search for a bigger site, casting his eyes south to a promising area of Holyrood Park, but he died in 1819 before the final move took the Garden in the opposite direction. A new site of just under 6ha in the Inverleith district of northern Edinburgh was purchased in 1820 from James Rocheid (who continued to live with his family in the splendid Inverleith House close to the western edge of the Garden). In those days much of Inverleith was parkland surrounded by nursery gardens. At the time of purchase, the new botanic garden – right next to the experimental garden of the Caledonian Horticultural Society (which later became the Royal Caledonian Horticultural Society, or simply 'The Caley') – contained a few trees and a kitchen garden but there were no other plants or garden features. The Leith Walk Garden contained many rare and interesting plants which could not be left behind; however, they were large and well-established. The newly appointed Regius Keeper, Robert Graham, was confronted with a major challenge: how to transfer the Living Collection through suburban streets to Inverleith. His Curator William McNab was not daunted. He invented a tree transplanting machine especially for the purpose and for the next two years set about the task of moving forest trees and flowering shrubs to their new home 2.4km away. It is a remarkable story best told by a contemporary commentator, Sir Henry Steuart, who gave his own first-hand account of the process a few years later.

below left William McNab, Principal Gardener 1810–1848. McNab is best known for his innovative management of the move from Leith Walk to Inverleith with the invention of a tree transplanting machine capable of moving mature trees. McNab was also an inspiring gardener, excellent teacher and talented plantsman with a passion for Cape heaths.

below middle Robert Graham, Regius Keeper 1819–1845. Graham supervised the move from Leith Walk to Inverleith. He also established the Tropical Palm House. A medical doctor, Graham was co-founder of the Botanical Society of Edinburgh (now Botanical Society of Scotland) with his successor and former pupil John Hutton Balfour.

below right Daniel Rutherford, Regius Keeper 1786–1819. A medical doctor, interested in chemistry and literature (Sir Walter Scott was his nephew), Rutherford is credited with the discovery of nitrogen gas. Perhaps his greatest gift to RBGE was appointing William McNab as Principal Gardener.

Master of invention

Moving the Garden from Leith Walk to Inverleith in 1820–1823, William McNab transported mature forest trees across town with the help of his transplanting machine and up to 12 horses. The story is told by Sir Henry Steuart in his *The Planter's Guide* (1828).

"The method adopted was to raise as great a mass or ball of earth as possible with the plants, and that was carefully matted up, in order to preserve it entire. The plants were then put upon a platform with four very low wheels, in an upright position, and transported about a mile and a half to the new Garden. In removing the trees, owing to the immense friction occasioned by the lowness of the wheels, ten and twelve horses were occasionally employed; so that the procession through the suburbs for many days, consisting of men, and horses, and waving boughs, presented a spectacle that was at once novel and imposing. The citizens of Edinburgh were surprised and delighted with the master of an Art, which seemed more powerful and persuasive than the strains of Orpheus, in drawing after it, along their streets, both grove and underwood of such majestic size.

"On arriving at the place of their new destination ... the trees and bushes were carefully planted. Numerous ropes, fastened pretty high from the ground, and extending from the stems to the distance of from twelve to four-and-twenty feet out, in the fashion of a well-pitched Bell-tent, pinned them to the spot with immovable firmness, so that injury from wind seemed altogether impossible. In this way, as may be easily conceived, little or no loss of plants could be sustained by the operation. The depth and richness of the soil; the sheltered site of the Garden, almost as low as the level of the sea; the steadfastness of the plants, in consequence of the fastenings; added to careful watering daily repeated, almost precluded contingency.

"... when the oldest of the trees had not been longer than a twelve month in the ground ... I was at the same time informed, that only one, or two at most, had died in the first season. On visiting the garden again in July 1827, I was both pleased and surprised to observe, that the more delicate plants, such as the Magnolia, the Perfumed Cherry, the Arbutus, etc had succeeded the best ... Of the Arbutus there is a noble specimen, supposed to be one of the largest in Britain.

"The ordinary Forest-Trees, on the other hand, such as the Lime, the Birch, and the Walnut, appeared by no means successful ... but they were placed in more exposed situations, and seemed less in possession of Protecting Properties. This conjecture was confirmed to me by the intelligent Mr McNab, who stated, among other things, that in the tallest of these Trees, which were from seven-and-thirty to three-and-forty feet high, the roots did not exceed three and a half or four feet in length; a style of root, as I observed to him, wholly inadequate to nourish or support plants of a far smaller size ... But I wish distinctly to repeat ... that I consider Dr Graham as beyond comparison the ablest, the most ingenious, and the most successful Horticultural Transplanter in Britain, or perhaps in Europe."

McNab's inventiveness was characteristic of the time. The world outside the Garden was changing rapidly as the Industrial Revolution gained pace in Scotland. Steam power speeded transport and transformed technology, enabling gardeners to create new environments in glass palaces for tender plants, as botanists explored further and faster round the world. New plants coincided with new ways of understanding the natural world. As always, the Garden responded by growing, this time upwards as well as outwards. Iconic architecture symbolised the excitement of the new era. The octagonal Tropical Palm House appeared in 1834, the tallest of its kind in Scotland at the time. The even larger Temperate Palm House opened in 1858. But there were other important, if less visible, signs of a new sense of purpose for the Living Collection.

Having secured the Palm House, in 1836 Regius Keeper Graham helped to found the Botanical Society of Edinburgh (now the Botanical Society of Scotland) along with his successor and former pupil John Hutton Balfour. Their aim, "to bring Botanists more

left The octagonal Palm House, built in 1834 at a cost of £1,500. The original construction featured a pyramid-shaped wooden roof and no surrounding lean-to glasshouses. Now called the Tropical Palm House, it was not at that time kept particularly warm and contained half-hardy palms, conifers and tree ferns. A list of 1870 shows examples of palms from eight countries including *Trachycarpus fortunei*, from north China, *Ceroxylon andicola*, from the Colombian Andes, and *Livistona australis* from Australia, as well as species of the conifer genera *Dacrydium* from South-East Asia to New Zealand and *Agathis* from the Philippine Islands to New Zealand.

below The tallest traditional glasshouse in Britain, the Temperate Palm House was built in 1858 at a cost of £6,000. In 1870 it housed plants such as *Arenga pinnata* and *Sabal umbraculifera* which had been brought from the Leith Walk Garden in 1822. Here also were the date palm, *Phoenix dactylifera*, and the talipot palm, *Corypha umbraculifera*, from South India and Ceylon. Apart from palms there were bananas, bamboos, sugar cane, allspice and pimento, along with figs, mahogany and much more.

immediately together", led to the foundation of a public Herbarium and Library, which merged with the collection of dried plant specimens previously held at the University and eventually moved to the Royal Caledonian Society's Exhibition Hall in 1863 (creating the basis for RBGE's world-famous Herbarium, for which a new building, designed for the purpose, was opened in 1964).

The changing status of botany also showed in an interesting rivalry between two outstanding candidates for the post of Regius Keeper when Graham died in 1845. John Hutton Balfour had to compete with Joseph Dalton Hooker and the competition was so strong the Town Council offered Balfour the Professorship of Medicine and Botany while the Crown offered Hooker the positions of Regius Professor of Botany, Keeper of the Garden and Queen's Botanist. Showing great determination, Balfour refused to accept the professorships without the royal titles and eventually won the combined posts. Hooker went on to become

Director of Kew in 1865.

Meanwhile James McNab, Curator of the Royal Caledonian Horticultural Society, took over from his father William as RBGE's Principal Gardener in 1849. A formidable partnership was forged between Balfour and McNab and – with the Palm House now seriously overgrown – history was about to repeat itself: the new Palm House would demand another major transplanting.

The construction of the new Palm House solved the problem of overcrowding under glass but the same problem was also occurring outdoors, with many trees and shrubs planted wherever space could be found. As a result the size of the Garden was almost doubled in 1864 – when it took over the garden of the Royal Caledonian Horticultural Society. This provided room and inspiration for James McNab's famous (sometimes infamous) first rock garden during the early 1870s. In roughly 30m by 10m McNab grew a wide range of alpine plants, spring bulbs and flowering annuals in thousands of individual stone compartments.

1

2

5

6

9

10

13

14

Museum Building, Curator's House and Lecture Theatre, 6 April 1909

Inverleith Row frontage, 6 April 1909

Pond, 1906

Arboretum from Palm House roof, June 1905

Pond c.1905–1907

Herbarium c.1905–1907

Palm house (Octagonal); photograph used to secure funding for new Palm House, c.1854

8 Tree lifting: *Quercus* on trolley; staff shot, c.1906

9 Staff group: gardeners with grass-mowers and pony, c.1904–1906

10 Staff group: foresters/gardeners by Glass Department with rakes, shears, hoes etc, c.1904–1906

11 Rock Garden showing the ornamental stairway to the viewpoint above the toolhouse, c.1894

12 Herbarium staff: Mr

Jeffrey etc, standing at Herbarium door; Clementina Traill (George Forrest's fiancée) on far right, 1905

13 Tree transplanting: final operation in RBGE; *Araucaria*, 1894

14 Tree transplanting: apparatus for removal and men at work, 14 March 1894

15 Surveying class with instruments, c.1894

15

Gardening under glass

Edinburgh's two grand old Palm Houses are dramatic landmarks in RBGE's long history of gardening tender plants under glass.

The iconic buildings are monuments to an age of human invention, discovery and determination. Horticultural imagination was fired by the Industrial Revolution. New technology provided the potential to construct artificial environments on a grand scale at a time when huge numbers of new plants were arriving in the Garden from warmer parts of the world. Lists of newly introduced species at RBGE closely reflect Britain's 19th century colonial and sea-trade interests: plants from Southern Africa (in particular Cape Province); plants from southern North America/Mexico, the West Indies, North America, the East Indies, China, Japan, Australia/New Zealand and India; and plants from the Mediterranean.

Regius Keeper Robert Graham was keen to build hothouses for this burgeoning non-hardy collection. Remarkably, although RBGE's annual budget was only £1,000, Graham persuaded the Government to donate £1,500 towards the construction of the first Palm House (now known as the Tropical Palm House). The octagonal building 18m high and 14m wide was the largest of its kind in Britain when it was built in 1834. (This was two years before the completion of the Great Conservatory of Chatsworth which inspired both Kew Palm House, built in 1848, and Crystal Palace, built in 1851.)

Glasshouses were not new at RBGE. As far back as 1711 George Preston, then Master of the Physic Garden, had built the Garden's first greenhouse, "therety four feet in length and sixteen foots in breadth" to preserve and cultivate "exotick or rare seeds". The move to Leith Walk in 1763 came with Treasury funding to build a greenhouse and two hothouses, and handwritten lists from 1774 and 1776 include 17 pages of "Tender Exotic Plants".

The great Victorian glass and iron buildings, however, were revolutionary in design, scale and heating technology. The octagonal Palm House (which predates Queen Victoria's coronation by three years) was heated at first partly by steam and partly by hot water but subsequently by hot water alone, using two cast-iron boilers. Plants grew quickly in the humid warmth – in fact the wine and sago palms (*Caryota urens* and *Metroxylon rumphii*) from Malaya grew so well that from time to tim they literally sent their leaves thro the roof.

By 1853 the next Regius Keepe John Hutton Balfour, was raising money to build another Palm Hou (now the Temperate Palm House, left). With £6,000 of public money construction on Robert Matheson' design started in 1856 and the bui was completed two years later. The new building measured 15.24m to top of the stonework, with each g dome 3.35m, giving a total height 21.95m. Palms were moved from t older Palm House, including at lea two specimens that had come fro the Stove House in Leith Walk: the Chinese fan palm (*Livistona chinens* and the royal palmetto (*Sabal umb culifera*) from the West Indies, whic weighed more than 7 tonnes and

measured 12.7m tall. Like his father before him, James McNab devised ingenious ways of repotting and moving these huge palms without accident or injury to plants or people.

For the following 160 years the Garden alternately restored and expanded glasshouses, re-inventing the purpose of the buildings as new plants swelled the Living Collection. In the 1870s a splendid Stove House provided heat and humidity for a wide range of economic, medicinal and decorative plants such as cinnamon, lychee, cloves, cotton and the massive coco de mer palm (*Lodoicea maldivica*) from the Seychelles. In 1908 a period of intense construction and renovation added a new tropical fern house, a temperate fern house and a heath house. By 1915 a new rhododendron house and two houses

devoted to alpines were completed in addition to a house for pitcher plants (*Sarracenia*) and one for bromeliads (species of *Billbergia, Cryptanthus, Tillandsia* and *Vriesea*) and there was new space for orchids (top right, p.50) and filmy ferns.

By the mid 1930s the Plant Houses built in 1832 bore little resemblance to the original structure. Over 150m long, the range of wooden framed glasshouses displayed succulents (middle and bottom right, p.50), orchids, ferns and stove plants. Corridors were lined with climbers, border plants and trees such as magnolias and birds of paradise (*Strelitzia reginae*). These glasshouses had fallen into disrepair by the 1960s (seen being demolished, above left) but, in good time for the Garden's tercentenary, they were replaced by

a magnificent new structure that was said to equal Chatsworth's Great Conservatory of 1837.

The Front Range (under construction and completed, top and bottom right), officially opened in 1967, was designed to include wide paths and an uninterrupted internal space made possible by suspending the houses from an external superstructure. The split-level structure of ten new glasshouses ranged across five climatic zones: from the dry heat of the Cactus and Succulent House (now the Arid Lands House) to the humid atmosphere of the Tropical Aquatic. The topsoil came from the construction site of the Forth Road Bridge.

The collection of tender plants continues to grow along with the glass structures to cultivate them. In 1978 Montane and Wet Tropics

houses were built for members of the ginger family (Zingiberaceae), the African violet family (Gesneriaceae) and tender rhododendrons (section *Vireya*). Behind the scenes, 'back-up houses' provide space for propagation, quarantine and research.

Glasshouses are constantly evolving but the historic legacy survives. At the turn of the 21st century the Temperate Palm House was restored and refurbished, opening up the central area, installing sandstone flooring and fitting traditional cast iron grids over the water-heated pipes. The Tropical Palm House, newly landscaped, remains at the centre, not only the oldest glasshouse in Edinburgh but still home to the Garden's oldest palm, *Sabal bermudana*, which was transported by horse from Leith Walk in 1822.

CALEDONIAN
HORTICULTURAL SOCIETY

left This Diploma certifi⟨⟩ depicts two founding fat⟨⟩ of the Caledonian Hortic⟨⟩ Society, 'The Caley' (later⟨⟩ Royal Caledonian Hortic⟨⟩ Society): Dr Andrew Dun⟨⟩ (left) and Dr Patrick Neill⟨⟩ Between them are the W⟨⟩ Garden conservatory (de⟨⟩ ished many years ago) a⟨⟩ Society's Show Hall.

below left The Caley Ha⟨⟩ today, now used for man⟨⟩ Garden functions. The Ca⟨⟩ was established in 1809⟨⟩ and acquired a 4ha site a⟨⟩ Inverleith in 1820–1821⟨⟩ RBGE was moving from⟨⟩ Leith Walk – to use as an⟨⟩ experimental garden. On⟨⟩ site the Caley constructe⟨⟩ glasshouses, a vinery, a st⟨⟩ house, a camellia house, ⟨⟩ conservatories, a propaga⟨⟩ house, a Winter Garden a⟨⟩ Exhibition Hall.

After getting into⟨⟩ financial difficulties in 18⟨⟩ the land was bought by t⟨⟩ Government and transfe⟨⟩ to RBGE. The Caley contir⟨⟩ today as a thriving hortic⟨⟩ society but no longer ow⟨⟩ any land.

opposite Inverleith Hou⟨⟩ on the west side of the Ga⟨⟩ was built for the Rocheid⟨⟩ ily in 1774, and was occup⟨⟩ by James Rocheid when R⟨⟩ arrived on the neighbour⟨⟩ site in 1820. The Garden⟨⟩ was expanded when Inve⟨⟩ Estate – including the ho⟨⟩ and surrounding land – w⟨⟩ acquired by the Town Cou⟨⟩ in 1877. The house, which⟨⟩ became the Gallery of Mo⟨⟩ Art displaying Henry Moo⟨⟩ sculptures on the lawn, is⟨⟩ RBGE's art and exhibition⟨⟩ gallery.

A question of security

The acquisition of Inverleith House and surrounding land created a problem for Balfour's successor, Alexander Dickson. When Dickson became Regius Keeper in 1880, the new site, now designated as the Arboretum (and funded by the city), was open seven days a week and separated from the rest of the (Crown administered) Garden by a high wall. As the Garden was closed on Sundays Dickson refused to open the connecting gates, maintaining that there was a risk to the Living Collection in having an unguarded entrance from the Arboretum when the main entrance to the Garden was guarded to prevent plant theft. Dickson stood his ground for two years in a debate with the Town Council and Government until the Treasury ended the standoff by instructing the Regius Keeper to open one of the three connecting gates with a member of staff on duty.

Further expansion followed in 1877 when the Garden acquired the policies around Inverleith House. The Arboretum of the future was at this point still separated by a wall from the rest of the Garden but almost 10,000 trees and shrubs were planted around the perimeter between 1879 and 1880. James McNab did not live to see this work. He died in 1878, ending a remarkable period of 68 years for the father and son team and a high point in the history of the Living Collection.

For Balfour, plants were an invaluable part of teaching; by now Edinburgh had become the largest botanical school in Britain. In the summer of 1878 his lectures were attended by 412 pupils, including students of medicine, science and pharmacy as well as general students. Demonstrations were also given in the glasshouses. In that one year more than 47,280 cut specimens were used to support teaching – more than 100 for each pupil.

However, plants were also for public pleasure six days a week (Sunday closure was to become a highly controversial issue). Visitors exploring the Garden of the late 19th century would have found winding paths through large lawns with collections of plants arranged either taxonomically or in other groupings such as British natives, variegated trees or ferns. Plants came from widely varying sources, a curious result of history, exploration and fashion. Many, like weeping trees in the former 'Caley' land, were very 'horticultural'. They originated mostly by chance, in gardens or in the wild, sometimes by hybridisation, and were then perpetuated as cultivars. Other plants were British and European natives or would have come from further east and west reflecting British trade routes and colonisation. Among the flowering shrubs there were a few new species of *Rhododendron* from India and China.

No one could have imagined how much all this was about to change.

An explosion of Rhododendron

Explosion may not be too strong a word. The influx of new plants arriving in the Garden from the turn of the 20th century certainly had an explosive effect on the Living Collection. Nowhere was this more dramatic than in the cultivation of spectacular new rhododendrons introduced from China – massive-leaved *Rhododendron sinogrande*, blood red flowered *R. delavayi* and the pale yellow blooms of *R. macabeanum*. The Garden had never seen such an invasion. In 1775 there had been three *Rhododendron* species at Leith Walk, in 1907 there were 135 species at Inverleith, by 1930 there were more than 400 and they kept on coming.

left *Rhododendron macabeanum*, collected by Frank Kingdon Ward in India in 1928. Rhododendrons changed the face of gardening for ever and Edinburgh was at the centre of their classification and cultivation – a position it holds to this day.

Man of vision

Isaac Bayley Balfour (1853–1922, pictured right) had close contact with the Garden from an early age. He was raised at 27 Inverleith Row, which backed onto RBGE, when his father was Regius Keeper. After studying at Edinburgh University he became its first Doctor of Science for a thesis inspired by his role as expedition botanist aboard HMS *Challenger* on the 'Transit of Venus' voyage to Rodriguez in 1874. On a further expedition, undertaken while Professor at Glasgow, he explored the botany and geology of Soqotra, an island situated at the far north of the Indian Ocean. From Glasgow he went to Oxford as Sherardian Professor of Botany before returning to Edinburgh in 1888 to become Queen's Botanist and Regius Keeper of RBGE. Sir Isaac Bayley Balfour was knighted in 1920.

An outstanding academic, visionary administrator and inspiring teacher, Balfour modernised the Garden, introducing horticultural apprenticeships and extending glasshouses and laboratories. Under Balfour RBGE led the introduction of new plants from China, changing the look of the Living Collection and establishing RBGE's international work on rhododendrons and other Sino-Himalayan flora.

During his tenure, on 1 April 1889, the Garden was transferred to the Commissioners of Her Majesty's Works and Public Buildings.

It is difficult to exaggerate the impact rhododendrons would have on both science and horticulture at RBGE. As Regius Keeper, Isaac Bayley Balfour was faced with the twin challenges of classifying a huge diversity of new and old species and finding space for them to grow. However, he played a much bigger part than that. Edinburgh was among the first to introduce these new species largely because of his foresight, enthusiasm and direction in the first place.

Bayley Balfour, the son of John Hutton Balfour, became Regius Keeper in 1888. The former plant explorer was to prove both an outstanding academic and a visionary administrator. He could also be an accomplished politician. His predecessor, Alexander Dickson, had been the first person to hold the position solely as a botanist. Balfour retained the post of Professor of Botany at the University of Edinburgh but successfully defeated a proposal to transfer ownership of the Garden to the University. That would have left the Arboretum (adjacent to the Garden but still separated by a high wall) under Crown control. As a result of Balfour's campaign, however, on 1 April 1889 the Garden was transferred to the care of the Commissioners of Her Majesty's Works and Public Buildings. From then on the Garden was open to the public seven days a week from dawn to dusk.

The following year the Treasury carried out an audit of the finances and functioning of the Garden and took a hard look at the Living Collection. The 1890 audit commended the herbaceous plant cultivation but criticised the indoor collections. As a result Balfour initiated a programme of restoration, rebuilding and restructuring both indoors and outdoors. The 185m long Herbaceous Border was designed in 1902 and the Beech Hedge behind it was planted in 1906. A year later McNab's Rock Garden featured unfavourably in Reginald Farrer's book *My Rock Garden*, leading to a complete reconstruction which took six years.

A golden age of plant collecting

Edinburgh was not unique. The influx of new plants from China was changing gardens throughout Britain and, eventually, across temperate regions of the world. The first three decades of the 20th century were to become known as a golden age of plant collecting. Edinburgh benefited in particular from close contact with explorers such as Forrest, Wilson, Kingdon Ward, Ludlow and Sherriff and others who introduced a vast array of newly discovered species from south-west China, Bhutan and Burma. Balfour, who

top The Herbaceous Border, photographed just a few years after its creation in 1902. At that time it was a mixed border of shrubs and evergreens; today it is composed of herbaceous plants only. The famous Beech Hedge at the back of the Border was added in 1906.

bottom left McNab's Rock Garden in 1890. Reginald Farrer, plant collector, garden writer and alpine plant enthusiast, described the stone compartments as "the Devil's Lapful", completing his criticism by adding that "the chaotic hideousness of the result is something to remember with shudders ever after".

bottom right Following Farrer's harsh criticism the Rock Garden was completely dismantled and reconstructed into its current form, a task that lasted from 1908 to 1914.

Forrest's 1919 Tsarang Expedition

A 2204 ? — Failed to Germinate 1921.
7 18926/18950 Catalapa sp. " " "
7 18937 Jasminum sp.
7 18899 Primula sp. nov.
7 18930 Magnolia sp. nov. aff. maclihmata — Failed to Germinate.
7 18896 Primula sp.
7 (not in packet when received)
7 18969 Meconopsis Henrica — Failed to Germinate
7 18965 Primula sp.
7 18936 Davidia sp. — Failed to Germinate

above Extract from the accessions book covering the years 1912–1923 and listing seeds from Forrest's 1919 expedition.

knew what it was like to collect seed in wild places, was an important mentor to the young George Forrest who began work as Assistant in the Garden's Herbarium in 1903.

All these discoveries were recorded in handwritten lists with detailed information about the plants and exactly where they were found. Such records are the bedrock of botanic gardens; the information they contain is what distinguishes botanic gardens from other gardens. In Balfour's time they were held in large ledgers called accession books and they reveal a fascinating story of exploration and introduction. They are witness to the massive influx of Chinese material that was to change the look of the Living Collection.

Volumes from the years 1912–1923 list donations of herbaceous plants from hundreds of individuals, botanic gardens and commercial nurseries. Botanic gardens, arboreta and museums included Amsterdam, Berlin, Calcutta, Dunedin, Portugal, Dijon, Strasbourg and Bucharest. Collectors, landowners and plant enthusiasts accounted for the greatest number and read like a 'Who's Who' of the plant world. Among them was Arthur Bulley from Cheshire, a cotton-broker in Liverpool and founder of the seed and nursery firm Bees Ltd. Bulley engaged Forrest, Kingdon Ward and Cooper to collect plants and the 26 consignments of plants he sent to RBGE

were mostly derived from these collectors. One such consignment was of 42 plants collected by R.E. Cooper from the Himalaya, Kulu and Lahaul. A similar volume for woody plants contains an equivalent range of donors and includes a rich list of genera such as maple (*Acer*), *Berberis*, birch (*Betula*), mountain ash or rowan (*Sorbus*), pine (*Pinus*), fir (*Abies*), oak (*Quercus*) and, of course, *Rhododendron*.

For Balfour there was now a formidable taxonomic task to be carried out. His classification system of the genus *Rhododendron* was essentially designed to cope quickly with the vast quantity of new plant material then being accessed. Sadly, Balfour died in 1922 before the revision task could be accomplished. But of course the work went on. After 34 years as Regius Keeper Balfour left a transformed and modernised Garden and scientific institution. His taxonomic work had established a valuable synergy between horticulture and science that would become a hallmark of the Garden for the next 100 years. In the more immediate future he had also opened the way for the Living Collection to grow beyond the Garden and across Scotland.

Classifying *Rhododendron*

In terms of classification, the genus *Rhododendron* was established by Linnaeus in 1753 when he described nine species.

right Portraits of gardeners 1910–1920.

Pioneers of the golden age

There were many plant collectors but six names appear on RBGE Garden plant labels with particular frequency:

Reginald Farrer (1880–1920, not depicted), from Clapham, Yorkshire, collected plants with W. Purdom in China from 1914 to 1916 and with E.H.M. Cox, from Glendoick near Dundee, in Upper Burma in 1919. He wrote many books about his travels, gardening and rock gardens.

George Forrest (1873–1932, above on horseback) was from Falkirk in Stirlingshire. He travelled to Australia and South Africa before becoming Assistant in the Herbarium at RBGE from 1903. He made seven plant collecting expeditions to Yunnan in 1904–1932 and introduced over 250 species of *Rhododendron*.

Frank Kingdon Ward (1885–1958, top left), from Manchester, was a schoolmaster in Shanghai from 1907. From 1909 to 1956 he collected in China,

Burma, Tibet and Thailand and later wrote many books about his exploits.

Frank Ludlow (1885–1972, second down on left), from Chelsea, London, was Vice-Principal of Sind College, Karachi and also opened a school in Tibet in 1923–1926. He collected plants and birds for the Natural History Museum, London. In 1928 he met George Sherriff. From 1942 to 1943 he was in charge of the British Mission in Lhasa, Tibet.

George Sherriff (1889–1967, third down on left), from Larbert in Stirlingshire, was the British Vice-Consul in Kashgar, Chinese Turkestan from 1928 to 1932. With Frank Ludlow he collected plants mostly in Bhutan and south-east Tibet from 1933 to 1949.

Ernest 'Chinese' Wilson (1876–1930, bottom left), from Chipping Campden, Gloucestershire, was a gardener at RBG Kew from 1897 to 1899 and collected plants for Veitch & Sons, nurserymen, from 1899 to 1902 and from 1903 to

1905. He returned to China on behalf of the Arnold Arboretum from 1906 to 1909 and again from 1910 to 1911; he then travelled to Japan in 1914 and 1917–1919 and the Far East, India and Africa from 1920 to 1921. He became Assistant Director of the Arnold Arboretum in 1919 and was Keeper from 1927 to 1930. He wrote numerous books about his travels and the plants he collected.

Between them the six collected and introduced numerous species of *Rhododendron*, *Primula*, *Gentiana*, *Meconopsis*, *Lilium*, *Viburnum*, *Sorbus*, *Picea*, *Abies* and more.

opposite top *Rhododendron orbiculare* ssp. *orbiculare* from Sichuan, China. The species was first introduced by E.H. Wilson in 1904.

bottom left *Rhododendron wardii* var. *wardii* grown from seed collected by Forrest on the Chienchuan-Mekong divide at 3,500m on 15 Ju 1922. The species is named after Kingdon Ward.

right *Rhododendron fortunei* ssp. *discolor*. This species was named in honour of Robert Fortune who introduced it in 1855.

above The Conifer Walk (situated between the Pond and Peat Walls and leading to the Woodland Garden) c.1905. Many of these conifers have now matured, leaving an open understorey which has recently been planted with ferns and other shade-loving herbaceous plants.

above View from the Temperate Palm House taken on 14 June 1911, looking west up the path leading to the Herbaceous Border on the right.

At the same time he created the closely related genus *Azalea*, into which he placed six species. By 1796 it was clear that the two genera should be merged into one and in 1834, George Don subdivided the genus *Rhododendron* into eight sections (recognised until 2004). With relatively few species known, their classification at this time was not particularly complex or controversial. However, all this was to change with the influx of hundreds of new species from the turn of the century.

Balfour's classification of *Rhododendron* was further refined by H.H. ('David') Davidian at Edinburgh, culminating during his retirement in his four-volume publication *The Rhododendron Species*, published between 1982 and 1995. Meanwhile, James Cullen, Deputy Keeper at Edinburgh, and David Chamberlain, a senior scientist specialising in temperate Asian plants, undertook a review of *Rhododendron* classification, based on sections and subsections and culminating in *A revision of Rhododendron 1, subgenus Rhododendron sections Rhododendron and Pogonanthum* by J. Cullen in 1980 and *A revision of Rhododendron 11, Subgenus Hymenanthes* by D.F. Chamberlain in 1982. In the late 1930s the then Regius Keeper, William Wright Smith, developed close ties with Chinese horticulturists and taxonomists. These ties were re-established in 1980, enabling the Garden to enhance its collections and undertake collaborative research on rhododendrons, which continues today.

How the trickle became a flood: rhododendrons in the Garden

Wonderfully diverse – from creeping alpines to soaring trees – the Garden's world-famous collection of rhododendrons now holds around half of all known species. The story begins with deceptive modesty 200 years ago.

Before 1900 rhododendrons played a small part in the Living Collection. In 1775 only three species were cultivated in Leith Walk – *Rhododendron ferrugineum*, the alpine rose from the Alps of Central Europe, and two species of North American origin, the rose bay (*R. maximum*) and spicily fragrant swamp honeysuckle (*R. viscosum*), both from the eastern USA.

A slow trickle of arrivals gradually increased at Inverleith. By 1787, lilac-pink *R. ponticum*, from Spain and Portugal, was in the collection. Others from further east followed before the end of the century. By 1813 *R. catawbiense* had reached the Garden from north-east America and *R. hirsutum* from Europe.

Old accession books catalogue the first consignments from the Indo-Himalayas, sent by people such as Francis Buchanan-Hamilton, a student of Hope who joined the East India Company and served in Nepal, and

The long tail of *Vireya* rhododendrons

In many ways RBGE's collection of *Rhododendron* section *Vireya* epitomises the very best that botanic gardens can offer – the plants produce a fine public display yet they are all wild collected, fully documented and of considerable scientific value, allowing the plants to be observed and described from flowering to fruiting (seldom possible during a short expedition). They have brought scientists and horticulturists close together through fieldwork, cultivation experimentation and scientific usage and they have also been used by artists and for student training.

Vireya rhododendrons typically grow in the montane forests of South-East Asia. A charismatic group of floriferous, colourful, often epiphytic plants, *Vireya* are defined as rhododendrons with scales whose seeds have a long tail at each end. This rather stark description is difficult to apply even when the plants have seeds. There are about 300 species with their most westerly distribution in Nepal and about 25 other species in mainland South-East Asia. Most, however, are from the islands of South-East Asia with, for instance, 22 species from Sumatra, 55 from Borneo and 164 from New Guinea, where they generally occur from 1,000 to 3,000m in altitude. Flower colour ranges from white and yellow through to orange, pink and red and size and habit range from small shrubs of less than 300mm to sprawling epiphytic scramblers of 5m or more.

Edinburgh's interest in this group derives from the Garden's general taxonomic interest in the genus *Rhododendron* and, in particular, from the work of George Argent. Argent became interested in them in 1977 and during the following 25 years built up the most comprehensive collection of *Vireya* species in the world. He finally published a definitive book on their description and classification (*Rhododendrons of subgenus Vireya*, published in 2006 by the Royal Horticultural Society in conjunction with the Royal Botanic Garden Edinburgh). Dr Argent has been one of the Garden's most consistent users of the Living Collection and has always invited horticultural staff to accompany him on his many field trips, resulting in a comprehensive, well-documented, well-grown collection of plants that is of enormous scientific value.

The Garden's main collection is held in one of the research glasshouses where the plants are grown in pots on benches, but a subset is cultivated in a more naturalistic setting in the Montane Tropics House. Here they luxuriate in the temperate, moist, humus-rich environment and show off their dazzling display of flowers. Their almost year-round flowering makes them valuable both for public display and for teaching.

below left top *Rhododendron lambianum* grown from seed collected on Gunung Alab, Sabah at about 1,800m on 15 February 1980 by George Argent.

below left bottom *Rhododendron phaeochitum* grown from seed collected in Marafunga, Papua New Guinea at 2,438m by Norman Crutwell in 1977.

below right George Argent with *Rhododendron anagalliflorum*, grown from seed collected near Bang on the way towards Mt Hagen in the West Highlands District of Papua New Guinea by Searle and Stanton in 1972.

Dr Nathaniel Wallich, Superintendent of Calcutta Botanic Garden from 1817 to 1846.

In the second half of the 19th century the flow increased. Hooker discovered thirty species in Sikkim and in 1850 a package of seed was forwarded from RBG Kew to Edinburgh with ten of 'his' species, including large-leaved *R. falconeri*, *R. grande* and *R. fulgens*, with its bright-scarlet flowers.

At this time the interior of China was effectively closed to most foreigners but a group of French missionaries with particular interest in botany – Delavay, David, Farges and Soulié – were allowed access. They collected seed and, mostly, herbarium material. While most of their material went to Paris many duplicate herbarium specimens came to Edinburgh. However, despite their discoveries, seed of only two species – the blood red flowered *R. delavayi* and the pink flowered *R. racemosum* – arrived here. The first Chinese species to be grown at the Garden was probably *R. fortunei*, around 1860.

By the end of the century over 300 species had been described – mostly from Europe, America and India, with some from the East Indies, China and Japan – but only about 30 species were in cultivation in Edinburgh. Some, such as *R. arboreum* and *R. hodgsonii*, were cultivated in glasshouses as they were thought to be too tender for outdoor cultivation. Those considered hardy were cultivated in the Copse and Woodland Garden and, after land around Inverleith House was acquired in 1877, in the Rhododendron Walk.

All that changed with the opening of China to foreigners following the Opium Wars. By 1930 there were over 400 species cultivated in the Garden and the majority of them had been collected between 1900 and 1930, including *R. sinogrande* with its massive leaves, discovered and introduced by Forrest in 1913, the large-leaved and exceptionally hardy *R. calophytum*, introduced by Wilson in 1904, and the large-leaved and pale yellow flowered *R. macabeanum*, collected by Kingdon Ward in 1928.

Since the reopening of China in 1980, wild origin rhododendrons are once again entering the Living Collection, notably from fieldwork expeditions such as the Chungtien, Lijiang and Dali Expedition in 1991 and the ten Gaoligong Shan Biotic Survey Expeditions undertaken from 1996 to 2006. The traffic, however, is not one-way. Species threatened in their native habitat now form the basis of repatriation programmes as plant material from RBGE is returned to China.

above left *Corylopsis sinensis* var. *calvescens*, one of the many other species of plant apart from *Rhododendron*, *Primula*, *Meconopsis* and *Lilium* collected in China in the early 20th century.

above right *Deutzia calycosa*, grown from seed collected on the Sichuan Expedition 1994, an example of a Chinese introduction from the late 20th century.

Growing across Scotland

Under the challenge of so many new discoveries, the Garden was running out of space. Isaac Bayley Balfour had been seeking new sites outside the Garden for his last ten years in office. The search was taken up with enthusiasm by his successor and former assistant lecturer, William Wright Smith. The new Regius Keeper, appointed in 1922, brought an ideal mixture of qualifications and experience to the job: an in-depth knowledge of the floras of India and Burma and a keen personal interest in forestry and timber supplies. Wright Smith's collaboration with the newly founded Forestry Commission led to a trial plot for exotic conifers at Inverleith – and a much bigger site on the west coast of Scotland.

left Benmore House, Argyll. With its high rainfall and dramatic topography the west coast of Scotland is far better suited to growing Sino-Himalayan plants than Edinburgh. Benmore Botanic Garden (initially called Younger Botanic Garden after the family who gave the estate to the nation) became RBGE's first Regional Garden in 1929.

above left View from c.1890s, across Strath Eck to the Formal Garden, Stable Courtyard buildings and, to the right, James Duncan's experimental sugar refinery.

above right Sir William Wright Smith was Regius Keeper from 1922 to 1956. He came to RBGE from the Royal Botanic Garden in Calcutta where he ran the herbarium, taking responsibility for the botanical survey of India in 1908. His keen interest in Sino-Himalayan plants helped to strengthen RBGE's links with great gardens and gardeners of the day.

The dry east coast coupled with the thin sandy soil of Edinburgh limited the true potential of Chinese and Himalayan plants. The mild, moist climate of Argyll – receiving about four times as much rain as the Edinburgh area – seemed good testing ground for the new plants. Trials began in 1925 on the 20ha Glenbranter Forest Reserve belonging to the Forestry Commission. Within four years a more appropriate site became available, closer to the ferry at Dunoon and framed by a remarkable collection of mature trees.

In 1929 the Garden acquired its first Specialist or Regional Garden: Younger Botanic Garden, now called Benmore Botanic Garden. This marked the beginning of a new expansion across Scotland, exploiting the full potential of a diverse climate, bringing new opportunities for scientific research and creative horticulture.

The Garden was fast gaining a reputation with both scientists and gardeners. Both Balfour's and Wright Smith's plant research focused on new Sino-Himalayan arrivals and as this material could be grown in the great gardens of the day Edinburgh became a Mecca for gardeners. RBGE prospered from the influence of notable plantsmen such as Colonel F.R.S. Balfour of

Dawyck in Peeblesshire, Major A. Dorrien-Smith of Tresco in the Scilly Isles and J.C. Williams of Caerhays in Cornwall.

New plant arrivals had a dramatic impact on the Garden. In many ways the structure of the landscape has remained the same. The Pond, Copse, Woodland Garden and Rock Garden – well known features of today's Garden – were established long before 1930. What was new was the huge range of newly discovered plants. Chinese discoveries made their way into the Rhododendron Walk winding round the Garden's central hill, under the shade of mature trees in the Copse and through the Woodland Garden. Rhododendrons even colonised a 'Rootery' of old stumps where dwarf species grew among *Rodgersia*, *Paeonia* and *Primula*. Species thought to be too tender to survive outdoors were grown under glass, including many now known to be perfectly hardy – such as *Rhododendron arboreum*.

Sir William Wright Smith (he was knighted in 1932) steered the Garden through a period of great change, including the special stress of the Second World War. Many of the structural developments – such as the expansion of the Rock Garden during the 1940s – were made to accommodate the influx of Sino-Himalayan plants.

opposite top The Rock Garden in 1923. Little seems to have changed in more than 70 years. However, the apparent continuity conceals a constant ebb and flow of individual species as they mature, die and are replaced.

bottom left The Pond in 1936. Then, as now, the Pond's margins were densely planted with moisture-loving plants such as *Primula* and *Astilbe* and the yew trees at the back (north) are just as they are today.

bottom right Rootery photographed on 30 May 1930. Sometimes also known as 'stumperies' today, these features were composed of tree stumps and planted with shade-loving herbaceous plants. The Garden is currently considering the creation of a new stumpery in the reconstructed Peat Walls to reduce the need for peat blocks.

1 Rose Garden: view from south-east corner looking towards the north-west 19 July 1924

2 Arranged Natural Order Beds at Edinburgh – view from the Palm House roof 16 April 1924

3 Forestry Nursery: laying on plants – distant view of men and method, 27 March 1925

4 Forestry Nursery: man using the 'Murthly Method' of growing forest seedlings, 27 March 1925

5 Plant House Range: Glass Department from west 28 March 1928

6 Orchid House: interior view 3 February 1926

7 Forestry Nursery: sowing operations. Preparing seed bed in drills by means of a roller 21 May 1925

8 Peatery: with Conifer background 16 May 1940

9 Plant propagation: practical demonstrating house of pans containing seeds in process of germination 8 April 1926

10 Plant propagation: recently erected glass corridor beside Herbaceous Potting Shed for containing the frames constantly used for propagation 8 April 1926

11 Plan of Demonstration Garden from the Garden guide of the 1970s

12 Arboretum: east side from roof of Palm House 16 April 1924

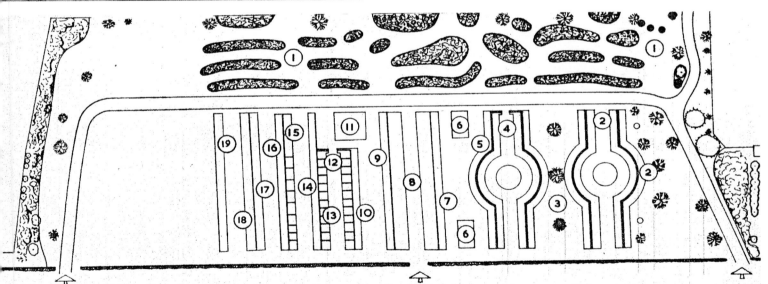

KEY TO DEMONSTRATION GARDEN EXHIBITS

1. Plant Family Relationships
2. Grass Collection
3. Seed Dispersal
4. Pollination
5. Poisonous and Irritant Plants
6. Forage Plants
7. Culinary or Sweet Herbs
8. Annual Borders
9. Chrysanthemums
10. Cultivated Gladioli
11. Climatological Station
12. Specimen Hedges and Ground Cover Plants
13. Antirrhinums
14. Michaelmas Daisies
15. Coniferous Hedges, Violas and Pansies
16. Garden Phlox
17. Lilium Species
18/19. Lilium Hybrids

Primula – classifying a specialist collection

Edinburgh's cool climate offered ideal growing conditions to many new species arriving from China and the Himalaya in the early 20th century. Dedicated research by Balfour and Wright Smith ensured that RBGE led the world in the classification of *Primula*.

At the 1913 RHS *Primula* conference, Balfour presented a paper noting that the number of known species had grown from 6 in the middle of the 18th century to 208 in 1905 – of which 54 were from India and 86 from China. Just 10 years later the number jumped to an astounding 300 species, of which 140 alone were from China. Now there are believed to be in excess of 500 species and subspecies.

The sheer number of these new discoveries rendered existing *Primula* classification outdated. Balfour picked up the challenge, adopting Pax and Knuth's approach of grouping the species into sections which he proceeded to double in number. Unfortunately he did not live long enough to refine his work but Wright Smith took up the task, increasing the number of sections to 32.

By the 1920s *Primula* were a conspicuous feature of the Garden and at least eight species, such as *P. alpicola*, *P. bulleyana* and *P. secundiflora* were displayed around the Pond. All eight originated in south-west China, Burma, Tibet and Bhutan.

Candelabra primulas became very popular in European gardens in the early to mid 20th century and Edinburgh was no exception.

Best-known was *P. pulverulenta* but there were at least seven further species, including *P. burmanica*, *P. beesiana* and *P. bulleyana*. Also in the Woodland Garden and Copse could be found *P. sikkimensis* – coming not just from Sikkim but also from Nepal, Tibet, Bhutan, Burma and Yunnan – as well as its giant counterpart *P. florindae* from south-east Tibet. Other species included *P. involucrata* and *P. yargongensis*, both ranging from the Himalayas to China, *P. rosea*, which grew well beside the Pond, *P. vialii* from western China and *P. edgeworthii* and *P. whitei*, which grew well in the Peat Garden.

Today's *Primula* collection is not quite as strong as it was in its heyday but the Garden still holds about 150 species and subspecies and numerous cultivars and hybrids. They are cultivated in the Copse, Woodland Garden, Rock Garden, Pond area and Rhododendron Walk.

below left *Primula pulverulenta* native to wet mountain meadows at 2,200–2,500m in west Sichuan thrives at Benmore.

right top *Primula forrestii* growing in Yunnan where it is found in limestone crevices at 3,000–3,200m. RBGE currently has eight wild origin accessions of this species.

right middle *Primula auriculata*, in north-east Turkey, where it grows in moist turf, wet flushes or by streams and lakes at 1,250–1,310m. RBGE currently has four wild origin accessions of this species which has a distribution in Turkey, Georgia and Armenia.

right bottom *Primula bulleyana*, growing in the Woodland Garden at Edinburgh. This accession was collected on the Alpine Garden Society China Expedition in 1994.

Wright Smith died in office in 1956 at the age of 81. By that time China was once more closed to botanists and gardeners from the west but an invaluable store of plants was in the safe keeping of the Living Collection.

Landmarks fit for the tercentenary

The riches of the Sino-Himalayan collections cast their spell over Harold Fletcher, who first came to the Garden in 1934 to fill a vacancy in the Museum Department. His early interest lay in the flora of Siam but long before he became Regius Keeper in 1956, Fletcher had succumbed to RBGE's *Primula* collection and was to publish extensive research on the genus.

New plants were coming into the Garden from a wider range of places than ever, and there was a new sense of purpose for all this living material. As the guidebook of the time says: "Not only was botanical and horticultural research undertaken in the Garden but horticultural staff were frequently asked to supply plant material for scientific research throughout the world. Much of this material might not have been of great ornamental value but every plant in the Living Collection served to increase the pool of genes on which progress in plant science depended."

During the 1960s the Garden was to be shaped by another strong professional partnership. Fletcher, who was Regius Keeper from 1956 to 1970, worked closely with Edward Kemp, RBGE Curator from 1950 to 1971, and together they made numerous additions and improvements to the still expanding site at Inverleith. Out of sight of the visiting public, the 4ha Nursery to the north of the Garden had been acquired in 1958 from a local nurseryman. The area to the north of the Beech Hedge, formerly a forestry nursery, became the Demonstration Garden in 1961 with the main aim of demonstrating evolution and the interrelationship of plant families. As the Garden grew in size so did the number and range of plants in the Living Collection.

Accession books were no longer large ledgers inscribed carefully by hand but were now composed of typed, pro-forma pages bound together. They still held fascinating details of the individuals, organisations and expeditions that supplied new plants. Then,

above left Harold Fletcher, Regius Keeper from 1956 to 1970. Building on the Garden's reputation in science and horticulture, Fletcher added new nursery space, new glasshouses and a purpose-built Library and Herbarium. In 1969, just before Fletcher retired, RBGE gained Logan Botanic Garden in Wigtownshire to grow tender southern hemisphere plants.

above right Edward Kemp, Curator from 1950 to 1971, worked closely with Fletcher to bring many improvements to the Garden. Perhaps his most notable achievement was the highly innovative new glasshouse range where not only was he responsible for spectacular plant displays but he also had considerable influence on the design and construction of the buildings.

above Library interior in the late 1970s. The 1960s were a period of innovation and modernisation and RBGE benefited with a new Library and Herbarium, opened by the Queen in 1964, and new display glasshouses, opened by Princess Margaret in 1967.

as now, the Garden had strong links with South-East Asia and an enormous amount of material came from non-RBGE expeditions or collectors. One intriguing example is the Admiral Paul Furse Expedition to Iran and Afghanistan in 1964 which produced (among many more) new species from genera such as *Bellevalia, Colchicum, Dionysia, Fritillaria, Primula, Scilla, Thalictrum* and *Tulipa*. The main difference from the 1912–1913 accession book is the significant increase in the number of RBGE-related donors, reflecting a growing range of botanical research and a wider reach across the world.

At the same time new landmarks were appearing in the Garden. A new Library and Herbarium was opened by Queen Elizabeth II on 29 June 1964. An innovative new glasshouse range vastly extended the cultivation of tender plants, including the giant Victoria lily (*Victoria amazonica*) which flowered in the Tropical Aquatic House with perfect

timing for the official royal opening on 25 October 1967. As RBGE moved towards its tercentenary two more significant changes took place. In 1969 Logan Botanic Garden became the second Regional Garden. With its uniquely mild and sheltered climate on the Rhinns of Galloway in south-west Scotland, Logan enabled the Living Collection to include material from the southern hemisphere. Less publicly visible, in the same year responsibility for RBGE was transferred to the Department of Agriculture and Fisheries for Scotland within the Scottish Office.

Behind the scenes

Like every Regius Keeper, Douglas Henderson was able to build on his predecessor's foundations. Henderson, a mycologist who had worked at RBGE for 20 years, directed the Garden from 1970 to 1987 through another era of rapid change (not least managing a bureaucratic change

opposite top The giant waterlily (*Victoria amazonica*) 1970, growing in what was then the Tropical Aquatic House of the new display glasshouses, now called the Plants and People House.

opposite middle Arboriculture in the Temperate Palm House, as practised in the early 1970s. Note the lack of protective clothing and the simple equipment.

opposite bottom The West Range in the 1960s located where the Alpine House now stands. While the house itself is old it was, in fact, bought second hand and erected in the 1950s. It appears in garden plans of 1957, but was demolished to make way for the Alpine House in 1974.

from civil service department to a new non-departmental body under a Board of Trustees). There were visible changes in the Living Collection – in 1974 a new Alpine House and trough garden offered a dazzling wealth of alpine gems and in 1979 staff tennis courts in the north-east corner were converted into the Silver Garden and soon became a favourite with the public (this area later became the Queen Mother's Memorial Garden). But the biggest changes happened out of public view.

A major review of the Garden's scientific programme in the early 1970s resulted in the recruitment of many new science staff and the generation of new research pro-grammes. The Garden's research glasshouses were built in 1974–1975, enabling science staff to hold material for experimentation and observation. There was no shortage of material to work on. New staff, new research programmes, new projects such as the *Flora of Bhutan* and *Flora of Arabia* and new botanical exploration led to a rapid expansion of the Living Collection and the number of specimens held in the Herbarium. Many science staff made excellent use of the new facilities, among them George Argent with his section *Vireya* rhododendrons.

This growth in research on living material brought changes in the Garden's plant-collecting strategy, and there was now a significant shift towards wild origin plants. The Garden's first *Collection Policy* (then called an *Acquisition Policy*) was published in 1981. Henderson's Assistant Keeper, James Cullen, worked with Curator Dick Shaw to champion the concept of wild origin collections, deliberately scaling back on the traditional sharing of seed from other botanic gardens (the seed list known as the *Indices Semina*). The launch of the *European Garden Flora*, another Cullen brainchild, brought horticultural taxonomists into the Herbarium with beneficial effects on the Living Collection. Henderson's tenure as Regius Keeper also coincided with the reopening of scientific links with China, starting with the Sino-British Cangshan Expedition in 1981.

Nearer to home, there was a different expedition into the hills. In 1979 Dawyck

Botanic Garden in the Scottish Borders became the third Regional Garden, bringing an invaluable collection of historic trees – and shelter for new discoveries from the east.

A call to action

Once again Chinese plants were colouring the landscape. For the first time in almost 50 years China was open to fieldwork expeditions from the west and once again

Edinburgh was to be a key player. During the early 1990s Edinburgh took part in a succession of collaborative botanical trips to north-west Yunnan, Kunming, Dali, Chungtien and Lijiang, establishing partnerships that have led to fieldwork in China almost every year since.

The Chinese Hillside grew out of the need to display a new influx of plants. The appointment of John Main as Curator in 1988 brought a new era of construction and a different way of displaying the Living Collection. Until then Chinese plants had been grown in different areas across the Garden without much logic in their display. The slope leading southward from Inverleith House – then planted with blocks of mostly hybrid rhododendrons without scientific or aesthetic value – offered an ideal location. Within three years the area was completely transformed with wild origin species growing in a more natural setting, arranged as they were found in the wild – lower altitude specimens at the foot, higher altitude at the top. A tumbling stream flowed into a pond overlooked by a Ting, or pavilion, winding paths led visitors up through the mountain

far left Douglas Henderson, Regius Keeper from 1970 to 1987. In a period of consolidation, Henderson widened public education and oversaw recruitment of new science staff, increasing systematic research and botanical exploration. During his time Dawyck Botanic Garden in Peeblesshire became the third Regional Garden, bringing an invaluable collection of historic trees.

left David Ingram, Regius Keeper from 1990 to 1998. Ingram was passionate about communicating plant science to a wide audience. He developed RBGE research and education, introduced an innovative science programme for schools, established the Friends of the Botanics membership scheme and the Botanics Trading Company. He also established new laboratories, new conservation and biodiversity research programmes.

Scottish Heath Garden

Before 1997 the area east of the Rock Garden displayed a collection of heather cultivars which had been popular with the public in the 1970s. But horticultural fashions change and the plants required rejuvenation. The solution was to retain the heather theme but to bring it up to date by creating a heath garden that represented a Scottish habitat.

While some non-native trees were kept for shade and shelter, the whole area was given over to native heath plants along with a peaty black bog and lochan or 'dubh lochan'. A ruined croft symbolised the social history of an environment where land was cleared of tenant or croft farmers to make way for sheep farming.

Now the area is also used to cultivate threatened Scottish species such as intermediate wintergreen (*Pyrola media*), woolly willow (*Salix lanata*) and Arran service tree (*Sorbus pseudofennica*). Interpretation panels tell an ethnobotanical story, explaining how people have used particular plants for thousands of years.

Remarkably, inside this urban garden it is impossible to see surrounding streets and buildings and there is a genuine feeling of being in a wild environment.

left View across the Scottish Heath Garden lochan, which was created in 1998 to replace the traditional Heather Garden of garden cultivars. Using wild origin native species, the Heath Garden attempted to simulate the landscape of the northern Highlands. It included ethnobotanical species, used for medicines, bedding and brewing, and a ruined croft, so typical of the Highland landscape today.

zones and plants quickly blended to create a vivid impression of a rural Chinese hillside.

This celebration of the wild world was part of a new determination to explain why people need plants. David Ingram became Regius Keeper in 1990, following John McNeill (Regius Keeper 1987–1989). In 1992 the first Earth Summit in Rio provided a dramatic impetus for the conservation of biodiversity through the kind of collaborative work RBGE was already engaged in. Ingram was passionate about

communicating an understanding of plants and science to a wide audience, particularly young people. While his eight-year tenure brought new focus on cutting edge science in new laboratory buildings, there was also an increase in practical conservation visible in the Garden, including the launch of the International Conifer Conservation Programme and the Scottish Rare Plants Project (now the Scottish Plants Project). The Garden became actively involved in the UK Biodiversity Action Plan and the

top Images of China. The Chinese Hillside, formally opened by Princess Anne in 1997, was constructed to display Chinese plants in a naturalistic way.

bottom Constructing the illusion. Creating pond, stream and bridges on the Chinese Hillside.

top left Stephen Blackmore, the 15th Regius Keeper, was appointed in 1999. Also passionate about communicating science to a wide audience, Blackmore is developing new ways to explain the value of the Living Collection with new gateway buildings at Edinburgh and Dawyck and an expanded RBGE membership scheme. He has added new research programmes, new education courses and international capacity building programmes. International capacity building programmes are not new to RBGE but have grown considerably in number and breadth since 2000.

middle left Field trip following a botanic garden development workshop at Nezahat Gőkyiğit Botanic Garden, Istanbul.

bottom left Surveying the site for the new Oman Botanic Garden.

below A bright spring display in the Alpine House where potted alpines are brought forward from the Alpine Nursery and plunge planted for a few weeks while they flower. The Garden has plans to add a new Alpine House where alpines will be grown naturalistically in a mountain setting of rocks and boulders.

Scottish Biodiversity Action Plan. Through the imaginative creation of more naturalistic landscapes – the Cryptogamic Garden in 1992 and the Scottish Heath Garden in 1997 – the Garden offered a new understanding of the complex relationships between people and the natural world.

The seeds of the future

By the turn of the 21st century the role of botanic gardens across the world was changing dramatically. A new sense of urgency directed the planning and planting of living collections as the diversity and survival of plants in their natural environment became increasingly threatened by the twin challenges of climate change and biodiversity loss.

Stephen Blackmore, the 15th Regius Keeper, arrived in Edinburgh in 1999 to take up office at a time of rapidly accelerating change. In Scotland 1999 was also the first year of a newly appointed Scottish Parliament, bringing an opportunity for RBGE to highlight its national and international value. Blackmore, like his predecessor, believed passionately in the importance of engaging the public and he understood that

the Garden's popularity as a major tourist attraction – drawing thousands of people to see living plants at all times of year – could be used to communicate urgent environmental messages.

This led to a newly structured membership scheme to increase public engagement and fundraising; new research programmes and investment in education; growing collaboration in conservation and research in over 40 countries, including Nepal, Turkey, Chile, Oman, Laos, Congo, Peru, China and countries in South-East Asia. As always new ideas made their way into visible displays in the Living Collection.

Today's Garden has come a long way from that small square near Holyrood Palace 340 years ago. The Living Collection is a rich amalgam of past and present; successive partnerships of Regius Keeper and Curator have bequeathed different legacies and distinctive landmarks while still retaining the best of what they inherited. From the moment Graham and McNab moved to Inverleith in 1822 until the present day, the Garden landscape and Living Collection have been continuously developed and improved.

With a little understanding of the history, visitors can read the landscape and the plants it contains. Parts of the external wall, for instance, show where the Garden was extended. The Caledonian Hall and East Gate Lodge are reminders that the south-east corner of the Garden was once the garden of the Royal Caledonian Horticultural Society. Inverleith House and the surrounding arboretum-style landscape are evidence of the private Georgian villa that once occupied the western part of the Garden. The gardenesque path and lawn layout are Victorian, the 'dotted' hollies and Herbaceous Border are classic Edwardian features and the wealth of rhododendrons and Sino-Himalayan plants are icons of the 'golden era' of Chinese and Himalayan plant introductions of the early 20th century.

The Rhododendron Walk was created so that visitors could see a little of everything in the Garden in just a short walk, starting at the West Gate and winding round Inverleith House. While the original design is not quite so obvious now, the area still contains some wonderful plants, such as the tree peonies *Paeonia delavayi* and *P. potaninii*, both collected in south-west China by George Forrest in the early 1930s. In the Woodland Garden, one of the most attractive areas of the Garden, the tree canopy is carefully manipulated to admit just the right amount of light to create the necessary ambience for ground flora growing in the shelter of mature trees. Here also are some of the Garden's most impressive large-leaved rhododendrons, mostly from subsections *Falconera* and *Grandia* – *R. rex*, *R. sinofalconeri*, *R. hodgsonii* and *R. macabeanun*

– some of which were collected by Kingdon Ward in 1928. In many ways these areas are little changed by time but there is new diversity among new plants grown from seed collected on recent field trips to China and Japan.

So to the present. The time traveller can continue round the Garden reading the influences and interests of each era and eventually arriving at the ecological plantings of today – not just in Edinburgh but also in the distinctive plant collections of Benmore, Logan and Dawyck. At Inverleith, the biggest visible change in the landscape of the new millennium has undoubtedly been the construction of the John Hope Gateway, officially opened by Queen Elizabeth on 12 July 2010. Celebrating one of the most

visionary Regius Keepers in the history of RBGE, the new building can be 'read' as a symbol of the environmental challenges and opportunities facing the early 21st century. Welcoming visitors to the Garden, the John Hope Gateway offers messages of hope and inspiration beautifully expressed in the surrounding planting of the new Biodiversity Garden.

Now growing across the four Gardens of RBGE – in four distinctly different parts of Scotland – the Living Collection has taken centuries to assemble and great skill to maintain. Today it includes plants from 161 countries, safeguarding an extraordinary sample of the world's botanical living inheritance – a legacy for the future for all to see and enjoy in the present.

above Welcoming visitors to one of the world's great botanic gardens, the John Hope Gateway and Biodiversity Garden engages public interest in the vital importance of plants. The newly landscaped building at the West Gate was opened by the Queen in 2010.

Benmore Botanic Garden

Surrounded by mountains, Benmore Botanic Garden is the largest, wettest and most dramatic of the four Gardens. The 49ha site is located along the Eachaig Valley on Argyll's Cowal Peninsula 5km north of Dunoon and 50km west of Glasgow across the Clyde estuary. It receives more than 2,600mm of rain a year – four times as much as the Edinburgh area. In recent years the maximum temperature has been 28.7°C and the minimum –8.1°C.

left Looking down into the Formal Garden at Benmore. Rhododendrons and conifers flourish in the wet, mild climate and seem at home in this large woodland garden. In the misty background are the slopes of Beinn Ruadh (left) and Meall Dubh (right).

above This fine specimen of *Rhododendron balfourianum*, grown from seed collected by George Forrest, thrives in the high rainfall at Benmore and looks at home in the informal woodland setting.

This mild, wet climate and spectacular setting combine to create a wonderful hillside garden on a grand scale. In addition to a historic collection of conifers and rhododendrons, new wild origin trees, shrubs and herbaceous plants from the world's temperate rainforest regions flourish among the mosses and ferns of Benmore on Scotland's wild west coast.

A wild and woodland garden

It was an extraordinary gift to RBGE. "A munificent gift" was how Regius Keeper Sir William Wright Smith described it when he first visited the site in 1925 after Sir Harry George Younger decided to donate his estate to the nation. In those days a metalled road took visitors along the Redwood Avenue to Benmore House and then, as now, the house was surrounded by mature conifers.

Benmore's plantations were the legacy of different owners of the estate who had turned the hillside green in just over 100 years. At the turn of the 19th century the old forested hunting grounds of the Dukes of Argyll had become treeless slopes grazed by sheep. Then, in 1831 George Ross Wilson founded the estate, built the first house and began to plant the first of Benmore's conifers, including newly discovered species from North America. In fairly quick succession each new owner made a different mark on the mountainside, perhaps most notably the American Piers Patrick, who is credited with planting the *Sequoiadendron giganteum* avenue in 1863, just ten years after the species was introduced to Britain. From 1870 James Duncan planted more than six million trees, adding greenhouses, a winter garden and a heated fernery high

on the hillside. In 1889 Henry J. Younger, the Edinburgh brewer, arrived and, together with his son Harry George, continued planting trees, shrubs and perennials, along with exotic flowering shrubs to the south of the Redwood Avenue (now the Younger Memorial Walk).

Harry Younger's offer came at a time when Wright Smith was already moving plants to an experimental site at Glenbranter, 5km north of Benmore. Wright Smith accepted the gift gratefully. Benmore was easier to reach and had better accommodation and a well-established landscape. In 1929 Younger Botanic Garden, as it was then known, became RBGE's first 'outstation', or 'satellite' garden (now known as a Regional Garden). The agreement was that the Forestry Commission would take control of the larger part of the 4,130ha

estate, leaving what was then 36ha of garden and arboretum to RBGE.

In the early years, Wright Smith and his deputy John McQueen Cowan sent thousands of recently propagated plants, especially rhododendrons, from Edinburgh. The climate suited them perfectly and young Sino-Himalayan plants looked more natural in this big hillside garden with space to grow as they would in the wild. Cowan's plan for the future recommended vistas of planted trees in the Eachaig valley and plantings above Benmore House (now used as an outdoor centre for Lothian Region schools). The official Garden Guide of the 1930s welcomed visitors to a "wild or woodland garden upon a magnificent scale ... a place where trees lend their canopy to make shelter and a home for well chosen aliens which add interest and an attractive richness

Sir Harry George Younger, of the Edinburgh brewing firm, gifted his estate with its house to the nation in 1929. Prior to that it was a fine country estate with forestry, agriculture, fishing and shooting interests.

top left Benmore House in the early 20th century.

top right Boy with two dolphins bronze fountain installed by James Duncan, owner of the estate from 1870 to 1889.

bottom left The Redwood Avenue in 1900, less than 40 years after the trees were planted by Piers Patrick.

bottom right Entrance hall of Benmore House showing Edwardian furnishing typical of a large Scottish estate house of the time.

of colour". Plants flourished but plans faded when the Garden suffered during the Second World War. An acute shortage of staff enabled Duncan's game cover of *Rhododendron ponticum* to take control of large areas of the hillside.

It wasn't until Dick Shaw became Benmore's first Curator in 1956 that staff were able to reassert control, knocking *R. ponticum* back with the help of newly powerful garden machinery. Shaw and his staff tackled a section of the Garden each year, clearing and replanting according to an agreed plan. Many plants were sent from Edinburgh and others were transferred from other parts of the Garden. Since then the small and highly skilled team of horticultural staff has continued to develop a distinctive Benmore blend of sensitivity and strength: essential qualities in the cultivation of a historically and botanically rich collection in such challenging terrain. The scale of the task grew during Douglas Henderson's

period as Regius Keeper (1970–1987) when a further 9.3ha of Benmore Hill and Glen Massan were added and today the Garden extends to 49ha, much of it on steeply rising ground.

Temperate rainforests of the world

In 1968, three years after Arthur Hall succeeded Dick Shaw, a hurricane struck central Scotland at 134 miles an hour. At Benmore hundreds of mature trees were flattened but the disaster also opened new opportunities in overcrowded woodland plantations. In the 1980s Benmore began to exploit the dramatic mountainous setting to mimic temperate rainforest regions around the world. With new RBGE expeditions exploring far to east and west there was both an opportunity and a need to create havens for plants threatened in their place of origin.

First the Tasmanian Ridge appeared above the Fernery in 1987. The Bhutanese Glade followed at the western end of Glen

top The Bhutanese Glade. Following an expedition to Bhutan in 1984, staff were inspired to create a microcosm of the Bhutanese mountains on a south-facing slope in Benmore. Planting started in 1989 with a second phase in 2001 adding a new generation of trees and shrubs as well as herbaceous perennials – all distributed up the slope much as they would be in Bhutan.

below The Chilean Rainforest is situated at the western extremity of the Garden on land formerly under commercial forestry. Following acquisition in 1993 the land was cleared and a network of paths created. The hillside has been planted with only wild origin Chilean plants, including Chile's seven threatened conifers as part of the International Conifer Conservation Programme project.

Massan a few years later after expeditions to Bhutan as part of the Garden's research in writing the *Flora of Bhutan*. The Chilean Rainforest began to emerge in 1997 with a more ambitious plan for newly acquired land at the far western extremity of the Garden. A 5ha site was planted with wild collected plants from Chile, focusing on threatened conifer species as part of the International Conifer Conservation Programme. Each one of these new developments involved heavy work clearing steep hillside of scrub woodland, self-sown conifers and ever-invasive *R. ponticum*.

In 2007 the Japanese Valley began to fill with newly collected material gathered on four field trips to Japan. Until 2003 RBGE had not undertaken any fieldwork in Japan for over 25 years and even then the only material brought back was conifers and ferns, mostly of unrecorded origin. Yet Japan, with its wealth of temperate plants, lies on the eastern extremity of the Sino-Himalayan vegetation zone and is therefore of long-term interest to the Garden. A wealth of material collected on these trips is now thriving on steep slopes near the Fernery.

Over time new plantings have matured. Curator Peter Baxter, who succeeded Arthur Hall in 1986, is now seeing the

top *Larix griffithii* growing in the Bhutanese Glade from seed collected by Ian Sinclair and David Long on 15 October 1984 between Paro and Chelai La, Bhutan, at 3,050m.

bottom The Chilean firebush (*Embothrium coccineum*) grown from seed collected on 15 February 1996 on the Instituto de Investigaciones Ecológicas Chiloé (IIECH) and RBGE Expedition, at 575m on the island of Chiloé, Chile.

first flowering of plants grown from seed he helped to collect on expeditions to Chile and Japan (he has most recently collaborated in expeditions to California). Such developments require long-term vision and patience. But on Benmore's hillside Tasmanian cedars (*Athrotaxis*) begin to offer shelter for younger southern beech (*Nothofagus*) and flowering shrubs. Bhutanese blue pine (*Pinus wallichiana*) climb towards larch (*Larix griffithiana*) and fir (*Abies densa*) at the top of a 156m glen. Among diverse Chilean rainforest plants, firebushes (*Embothrium coccineum*) and southern beeches wind their way up the hillside to maturing monkey puzzles (*Araucaria araucana*) which will one day dominate the view at the top. Young Japanese maples (*Acer rufinerve* and *A. capillipes*) are among the seedlings settling into the recently planted valley.

The collection of wild origin material is designed to be of scientific, conservation and education potential yet it also enhances the landscape. Perhaps more remarkably, these exotic plants look as if they belong there. Benmore's geographic areas provide a natural setting for plants from the other side of the world thriving among native ferns and mosses. In fact ferns, lichens and mosses are everywhere. While Britain might not be rich in native flowering plants, the Atlantic coast plays host to a diverse and internationally significant bryophyte flora often (but not always) found in association with natural hazel woods (*Corylus avellana*). Benmore's clean air and high rainfall make it an ideal place for mosses and liverworts to thrive in a great diversity of microclimates (rocks, rotting wood, pathways and stream-sides).

Since Sir William Wright Smith, successive Regius Keepers and their Curators have continued to make their mark on the Garden. Over the years, collections of specific genera have been cultivated and expanded at Benmore but the main emphasis has always been on Sino-Himalayan plants. Climate, topography and history have all influenced Benmore's great legacy of plants. The Living Collection also reflects major research interests of past and present, particularly in the mature conifers and rhododendrons growing much as they would in their native natural environment.

Benmore Fernery

In the summer of 2008 work began to restore James Duncan's 19th century fernery to its former glory. It was built in 1874 at the height of the Victorian fern craze and the construction must have been a major feat, creating two storeys of stone at the top of

below left Mosses flourish in the clean air and high rainfall at Benmore, often carpeting the ground in thick mounds.

below right Cones of the Korean fir (*Abies koreana*) on a tree grown from seed collected on the west slope of Hallasan, Republic of Korea, at 1,300–1,900m on 28 September 1976.

a narrow gulley without the benefit of heavy lifting machinery. Even with today's tools, restoring the derelict building was a remarkable achievement. The Benmore Fernery reopened in 2009 filled with a superb collection of ferns.

Hillside rhododendron collections

In the heart of the Garden grow the heart of Benmore's collections: fine mature rhododendrons. Where mature tree plantings provide shade and shelter, large-leaved species in subsections *Falconera* and *Grandia*

are to be found, such as *R. arizelum*, from Yunnan and northern Burma, with cinnamon-coloured indumentum beneath its leaves and creamy yellow flowers. Spectacular *R. grande*, which has white flowers with a purple blotch, is from east Nepal, Sikkim and Bhutan. In more open areas, species in subsection *Triflora* can be found: *R. davidsonianum* from south-west China introduced by E.H. Wilson in 1908 and *R. oreotrephes*, from south-west China, Burma and Tibet, introduced by George Forrest in 1913.

top left The ruined Fernery, first built by James Duncan in 1874, prior to the extraordinary task of reconstruction which started in 2008 and was opened to the public in 2009.

top right Construction underway. All remaining metalwork was removed along with accumulated debris, brambles and seedling trees. The walls were repaired and repointed before new glazing bars and glass were added.

bottom left Some elements of the interior landscape were intact, such as the grotto, but the entrance vault and steps had collapsed. The interior

is now resplendent with beds, paths, appropriate Victorian fern seat and, of course, a dense planting of over 70 species of ferns.

bottom right Most of the plants were propagated and grown on in the fern growing facilities at Edinburgh but some, such as this tall tree fern (*Dicksonia squarrosa*) were transplanted from the display houses.

opposite *Rhododendron degronianum* ssp. *yakushimanum* var. *yakushimanum* from Japan puts on a spectacular display.

Working together

International collaboration is a vital part of today's conservation and botanical research. Benmore is a store of living material gathered on important recent expeditions, among them the Instituto de Investigaciones Ecológicas Chiloé and RBGE Expedition to Chile in 1996 which collected 459 accessions. Two years later the Universidad de Chile and RBGE Expedition to southern Chile collected 281 accessions. There have been four trips to Japan, including the 2003 expedition which collected 303 accessions and the 2005 trip which collected 304 accessions, all undertaken in collaboration with colleagues from Japanese institutions.

The Garden travels east as well as west with numerous expeditions to China and the Himalayas. There were trips to Bhutan in the 1980s and, between 1996 and 2006, ten trips to the Gaoligong Shan in south-west China in collaboration with the California Academy of Sciences and the Chinese Academy of Sciences which brought back a total of 606 accessions.

Redwood Avenue

Planted in about 1863, 49 of the original 50 giant redwood (*Sequoiadendron giganteum*) trees survive and have reached a height of over 50m. The Avenue was originally a metalled road leading to Benmore House but, to protect the tree roots, was grassed over in 1973.

Running parallel to the Avenue on the south side, the Younger Memorial Walk was planted in 1916 to commemorate H.G. Younger's gift of the estate to the nation. The Walk is like 'Benmore in miniature' as it includes a little of everything the Garden has to offer.

Pond

Created in the days of Piers Patrick and adorned with a bronze fountain in the time of James Duncan, the Pond today is richly planted with naturalistic groupings of *Hosta*, *Rodgersia*, *Gentiana*, *Arisaema* and *Paris*. In autumn the centre is dominated by a spectacular *Cercidiphyllum japonicum*.

Fir Bank

Descending down from the Viewpoint in an easterly direction gives fine views of the Formal Garden and borrowed landscape beyond and up to the summit of Benmore Hill. As the name implies, this area is rich in species of fir (*Abies*) with mature specimens of giant fir (*Abies grandis*) and noble fir (*A. procera*). There are also younger, wild collected species in this area, such as Japanese fir (*A. firma*) and Nikko fir

(*A. homolepis*), both from Japan. Around the edge of the area there is a mixture of rhododendrons (including *R. sanguineum* and *R. haematodes*, both with red flowers) and *Sorbus* species (such as *S. decora*, *S. commixta* and *S. monbeigii*).

Like all great gardens Benmore will never be finished but will change and evolve under new Curators. Recent additions have seen the creation of the Courtyard Gallery, restoration of the Fernery and new, innovative plantings. It is hoped that in the not too distant future the Golden Gates will be restored and surrounding area landscaped, a new visitor centre built and the Formal Garden redesigned.

above Giant Redwood Avenue (*Sequoiadendron giganteum*), planted in 1863, soon after their discovery in California during the 1849 Gold Rush and their introduction into Britain in 1853. These trees, probably grown from seed introduced into Britain by the Cornish plant explorer William Lobb, are now over 50m tall and form the best giant redwood avenue in Britain.

Logan Botanic Garden

A magical surprise for visitors, Scotland's most exotic garden is set among farmland on the Rhinns of Galloway 14km south of Stranraer. Logan Botanic Garden is RBGE's smallest garden but the 11ha site is filled with a spectacular array of tender plants, many of them from the southern hemisphere. In a unique microclimate warmed by the Gulf Stream, the average annual rainfall is just under 1,000mm and the most extreme temperatures recorded in recent years are 27.7°C and –9.6°C, but on average the lowest temperature is only –5°C.

left Visitors enjoy the sub-tropical illusion created by cabbage palms (*Cordyline australis*), agapanthus and red hot pokers (*Kniphofia* spp) in Logan's Walled Garden.

above The Walled Garden, at the heart of the Garden, is a special mixture of architectural features – ponds, walls and terraces – and plantings of, mostly, southern hemisphere plants. The white cottage was once the Curator's house but is now a 'Discovery Centre' and Gallery – displaying educational information and exhibitions.

The Living Collection here displays plants from central and southern Asia, South Africa, Australasia and other regions of the world with a Mediterranean type climate. Elsewhere in Britain they would need the protection of glasshouses but in Logan's Walled and Woodland Gardens they thrive in the open.

Transforming an old-fashioned Scottish garden

"It was just an old-fashioned Scottish garden in years gone by, where everything grew side by side, vegetables, fruit and flowers and it is a very ancient garden …" In 1927, in an article for the Royal Horticultural Society journal, Kenneth McDouall eloquently described the historic garden his family had owned for centuries.

The McDouall brothers (Kenneth and

Douglas) completed the transformation from traditional kitchen garden to sub-tropical fantasy, but it was their mother Agnes who first introduced exotic plants to the old walled garden. When Agnes Buchan-Hepburn from the East Lothian estate Smeaton Hepburn married James McDouall in 1869 she brought with her a passion for gardening and some of her favourite plants. While her husband continued to improve the estate that had been in the family since 1295, Agnes planted the first *Eucalyptus* and began to experiment with other exotic plants. She passed her love of gardening to her sons, who not only tended the site but also travelled to collect seeds and plants as well as subscribing to expeditions of renowned plant hunters of the day: Reginald Farrer, George Forrest and Ernest Wilson. The McDoualls were particularly proud of their Canary Island

Echium plants, probably the first to be grown successfully in Scotland, raised from seed sent by Major A. Dorrien-Smith of the world famous Tresco Garden in Scilly.

Today's entrance to the Garden is lined by Chusan palms (*Trachycarpus fortunei*) planted by the McDouall brothers. The Walled Garden, resplendent with its tree ferns (*Dicksonia antarctica*) and cabbage palms (*Cordyline australis*) is still a magical surprise for visitors walking through the unassuming gateway in the wall.

After Kenneth McDouall died in 1945 Logan changed hands and, some years later, a trust was created to protect the estate's future. One of the first trustees was RBGE Regius Keeper Harold Fletcher. With rising maintenance costs, the trustees tried to sell the estate but to no avail. In 1969 a 6ha parcel of Logan estate was given to the nation – and

the Walled and Woodland Garden became RBGE's second 'specialist garden'. Logan estate's Head Gardener Martin Colledge, who had trained as a horticulturist at RBGE, became Logan's first Curator. The unique climate and collection of plants brought a wonderful opportunity to expand the RBGE Living Collection in an exciting direction, adding a new dimension to the store of material cultivated in the wilder, wetter setting of Benmore in Argyll and the colder, drier urban environment of Inverleith in Edinburgh.

Under RBGE, Logan's own collection of plants also expanded and developed enormously with a special focus on clearly designated groups. These included the historic Peat Walls (created by the McDouall brothers), Australasian Woodland, Tasmanian Creek, *Rhododendron* subsection *Maddenia*,

left Kenneth (left) and Douglas McDouall by the Verandah, *c*.1940. These remarkable brothers created a unique garden, full of botanical interest, still the basis of Logan today.

right Agnes Buchan-Hepburn, *c*.1865. The foundation of the present Garden stems from 1869 when James McDouall married Agnes, from Smeaton in East Lothian. She brought with her numerous plants and a great passion for gardening which she passed on to her sons Kenneth and Douglas.

right The Tasmanian Creek at Logan. Inspiration came when former Curator Barry Unwin was visiting Australia and New Zealand. The area, fed by a small stream, was developed with naturalistic planting from 2003 onwards. Numerous wild sourced tree ferns (*Dicksonia*) and blue gum (*Eucalyptus*) were planted and a viewing platform was constructed, taking advantage of the existing topography so that visitors could look down on the Creek.

southern hemisphere ferns, tender climbers, Chilean species, South African monocotyledonous plants and specific plantings for the International Conifer Conservation Programme. Added to this rich diversity there tender perennials bring a summer blaze to the Walled Garden and, among Scottish rare plants, species from coastal habitats.

Despite being one of the mildest gardens in Britain, Logan is still at the mercy of Atlantic gales and relies on protective shelter-belts. When RBGE took ownership the shelter was thought to be too small and in too poor a condition for the long-term well-being of the Garden. In 1976 a further 4ha of open land, south-west of the Woodland Garden, was acquired from the estate. This area, known as Deer Hill, allowed the creation of a windbreak and continuous planting until the mid 1980s would ensure protection from south-west winds that can topple trees and scorch tender plants.

A new era of development

When Martin Colledge retired in 1988 his position was taken by Barry Unwin,

A historic experiment

The Peat Walls contain a fine selection of Ericaceous and other peat-loving species such as *Rhododendron charitopes* ssp. *tsangpoense* and *R. trichostomum,* both of which grow to less than one metre tall. These are followed in summer by several species of blue poppy (*Meconopsis*) and *Primula*. These Peat Walls date back to the mid 1920s, when the McDouall brothers pioneered the technique of growing plants

preferring moist, acidic conditions in raised peat beds. The Peat Walls have been refurbished at least three times since Logan became part of RBGE, the most comprehensive overhaul being undertaken in the late 1980s when the whole area was sterilised to eradicate ivy-leaved toadflax (*Cymbalaria muralis*), which had become a bad weed, and grassed down for a year before starting again. At this time they were expanded to both sides of

the path along the boundary wall. In the winter of 2002–2003 they were completely rebuilt again before replanting in the spring with low-growing calcifuges such as species of *Rhododendron*. When the Peat Walls were first designed there was no concept of the need to conserve peatlands. Now, however, RBGE is very concerned about the use of peat and the most recent redevelopments have used as little as possible sourced from a local, low-grade site.

left The original Peat Walls *c*.1930, a few years after their construction.

middle Peat Walls one year after reconstruction.

right Plants, such as species of *Meconopsis*, orchid (*Dactylorhiza*) and *Rodgersia*, growing vigorously two years after reconstruction.

top right Former Curator at Logan Barry Unwin collecting seeds on the slopes of Mt Wellington, Tasmania, 2005.

bottom right A young plant of the alpine ash (*Eucalyptus delegatensis*), grown from seed collected in south-eastern Australia in 2005, thriving in the Tasmanian Creek.

who had been Supervisor at Logan for many years. Unwin was a superb practical gardener with a huge plant knowledge and a keen eye for detail. He was particularly interested in introducing even better documented, wild origin plants into the Living Collection. His fieldwork in New Zealand and Tasmania in 1996 produced a rich new source of material for the Castle Woodland and inspiration for imaginative new developments.

The Tasmanian Creek at Logan was inspired by Unwin's first-hand experience of tree ferns and eucalyptus growing together. Back at Logan work began during the winter of 2002–2003 to recreate a naturalistic Tasmanian habitat in the West Woodland. The area had been cleared of mature beech, sycamore and a scrub of the ubiquitous *Rhododendron ponticum*. Snowdrops were moved

A sub-tropical fantasy

The focal point of the Walled Garden is the Water Garden with adjacent tree ferns (*Dicksonia antarctica*) and cabbage palms (*Cordyline australis*), now 100 years old. The pond contains a range of waterlilies, and the surrounding beds include bold plantings of arum lily (*Zantedeschia aethiopica*), angel's fishing rod (*Dierama pulcherrimum*), green false hellebore (*Veratrum viride*) and *Ligularia veitchiana*. The north wall border hosts a fine collection of *Rhododendron* subsection *Maddenia* species and hybrids, such as *R. burmanicum*, *R. crassum* and *R. x fragrantissimum*. These sweet-smelling rhododendrons are not hardy in any other part of Scotland but are ideally suited to Logan's mild climate and therefore demonstrate very well the value of RBGE's Regional Gardens. The magnificent specimens of *Dicksonia antarctica* in the Tree Fern Grove are over 100 years old and are believed to have been obtained from the collection established by Joseph Paxton in the Crystal Palace in the late 1850s. Recently planted species in this area include *Grevillea rosmarinifolia*, *Lobelia tupa* and *Euryops pectinatus*.

below Old tree ferns (*Dicksonia antarctica*) thrive in the mild climate at Logan and add character to the Walled Garden.

bottom left Underplantings of the fern *Blechnum tabulare* and the bromeliad *Fascicularia bicolor* ssp. *canaliculata* continue southern hemisphere theme.

bottom right Arum lilies (*Zantedeschia aethiopica*), of borderline hardiness in many parts Britain, are completely hardy at Logan.

100

elsewhere in the Garden and a 90m length of *Griselinia littoralis* hedge was planted to provide shelter. Tree ferns were to be the main feature and the ground was enriched with many loads of compost before 200 young *Dicksonia antarctica* specimens were planted. A number of different eucalyptus species were also planted to give dappled shade and shelter.

Further fieldwork in 2005 to Tasmania, New Zealand and mainland Australia gathered over 550 accessions of seed – this choice selection included species of blue gum (*Eucalyptus*), bottle brush (*Callistemon*), *Brachyglottis*, tea tree (*Leptospermum*) and daisy bush (*Olearia*).

At the same time, interest had been growing in rare and threatened plants much closer to home. In May 2002, an interpretive planting of Scottish coastal plants 'Wonder at Your Feet' opened at Logan, linking with RBGE's Scottish Rare Plants Project. The planting recreated a natural coastal habitat with zones ranging from high tide line through to the upper cliffs, allowing visitors to 'walk through' a coastal transect and learn more about some of the rarest and most characteristic native plants in the south-west of Scotland. In 2010 these plants were moved and redesigned into a more naturalistic setting, using existing outcropping rocks and adding seaside artefacts including old ropes and a lobster pot.

New expeditions continue to bring exotic plants to the Garden. Successive Curators have enjoyed exploring Logan's potential to grow plants not normally able to survive outside in Britain. Recent plantings include tree ferns and other fern species new to the Living Collection, such as soft tree fern (*Cyathea smithii*) from New Zealand, *Thyrsopteris elegans* from Juan Fernandez Islands, *Lophosoria quadripinnata* from Chile and mountain blechnum (*Blechnum tabulare*) from South Africa. In the Walled Garden, the Rock Gulley has recently been replanted following inspiration gained during fieldwork in the Atacama Desert, with plants from northern Chile such as giant oxalis (*Oxalis gigantea*), *Cistanthe grandiflora* and *Dunalia australis*.

above Visitors arriving at the Garden are immediately confronted by the Chusan Palm Avenue (*Trachycarpus fortunei*) which gives a flavour of things to come in the Walled Garden and beyond. It is named after Robert Fortune, a Berwickshire explorer who travelled widely in China in the mid 19th century.

Logan Viewpoint

The Viewpoint shelter, erected in 1976, at the top of the South Woodland offers fine views over the Garden to the distant Galloway hills. The walk to the Viewpoint from the viewing platform includes numerous recently collected species from Chile. Prior to the Chilean Rainforest project at Benmore, where temperate species from central Chile were grown on a large scale and in as natural a condition as possible, Logan was the main site for RBGE's Chilean plants. This policy has been reassessed and Logan now concentrates on growing relatively small numbers of a wide range of more northerly species. Examples here include *Lomatia ferruginea,* which has fern-like leaves and red-brown velvety stems, *Amomyrtus meli,* a myrtle with fragrant white flowers and hard wood that in Chile is used for making tool handles, and *Rhaphithamnus venustus,* a spiny plant with blue berries which is threatened in its native habitat.

The aptly named Chilean firebush (*Embothrium coccineum*) can also be found here along with the evergreen shrub *Desfontainia spinosa* with its holly-like leaves and waxy red tubular flowers.

left *Azara serrata:* grown from seed collected on 10 March 1998 in the Malleco Province, Chile on the margin of a small river with *Podocarpus salignus, Austrocedrus chilensis* and *Desfontainia spinosa* during the Universidad de Chile and RBGE expedition to southern Chile.

above The giant leaves of rhubarb-like *Gunnera manicata* from the swamps of Brazil grow particularly luxuriantly at Logan where the Gunnera Bog, undoubtedly the most impressive in Scotland, becomes an almost impenetrable forest in mid summer.

Richard Baines, who became Curator following Unwin's retirement in 2007, has a special passion for palms. Keen to experiment with outdoor cultivation, he has added ten new species of palms, including *Juania australis,* a slow-growing threatened species from Juan Fernandez Islands, the pindo palm or jelly palm (*Butia capitata*) from Brazil and Uruguay, the Mexican blue palm (*Brahea armata*) and the threatened Brazilian mountain coconut (*Parajubaea torralyii*) which is endemic to Bolivia.

Logan today remains the exotic plantsman's garden established by the McDouall brothers in the early part of the 20th century. In recent years, the development of the Woodland Garden with its more naturalistic design has added a different dimension. Rich plantings, a mild maritime setting and its idiosyncratic design all combine to make it a very special garden, quite unlike any other in Britain. It has a most remarkable, diverse and well-documented collection of wild origin southern hemisphere plants that is undoubtedly the richest in the UK.

Avenue of Chusan palms

The main botanical feature greeting the visitor entering the Garden is the avenue of Chusan palms (*Trachycarpus fortunei*). This has been extended in recent years with more palms planted by RBGE's students during a visit to gain work experience.

Gunnera Bog

As the name suggests, the Gunnera Bog is dominated by an extensive planting of *Gunnera manicata* from Brazil. This dramatic plant amazes Logan's visitors with its enormous leaves that can grow up to 1.5m across. Walking within the Bog, with the leaves soaring above head height, is like being in a different world. However, after the first frost, the massive structures collapse completely.

Castle Woodland

The Castle Woodland is situated on the west side of the Walled Garden and is dominated by the remains of Castle Balzieland, a medieval residence that burnt down around 1500. The area immediately adjacent to the wall is devoted to *Rhododendron* subsection *Maddenia* species which, before RBGE had acquired Logan, were grown in a special conservatory at Edinburgh as they are not fully hardy.

A blaze of colour

In summer the central formal borders display a fine array of half-hardy plants such as species of *Salvia*, *Geranium*, *Argyranthemum*, *Gazania*, *Felicia* and *Osteospermum*, which are a well-loved feature of Logan, while the surrounding beds contain mostly herbaceous plants. Both produce a dazzling display of colour in the second half of summer. The walls behind are richly planted with species of *Clematis*, *Rosa* and *Actinidia*.

top left Richard Baines, Curator at Logan since 2007, collecting seeds in northern Chile to add to the collection at Logan.

top right The Lower South Woodland features species of fern and *Eucalyptus* typical of Australian forests.

below Half-hardy bedding plants such as this cultivar of *Gazania* are a special feature of Logan's Walled Garden. Grown in their thousands from cuttings each year, these plants are planted in the borders in April and May where they create a riot of colour from late July to October.

Dawyck Botanic Garden

On the upper reaches of the River Tweed in the Scottish Borders, Dawyck Botanic Garden is RBGE's coldest garden by far. Winter temperatures can be as low as -19.8°C and the growing season is short. With summer temperatures also slightly higher than Scotland's average – they have reached almost 30°C in recent years – its climate could be described as almost continental.

left Dawyck benefits from the fine borrowed landscape of the rolling Borders hills and 300 years of tree planting by three families – the Veitches, Naesmyths and Balfours.

left Conifers thrive at Dawyck and some of the earliest planted specimens are now of massive size. This Douglas fir (*Pseudotsuga menziesii*), one of the first to be planted in Scotland, dwarfs the Dutch Bridge, commissioned by Sir John Naesmyth in 1856.

right A touch of formality at Dawyck: Sir John Naesmyth added stone urns, steps, seats, balustrades and grass terraces to Dawyck's otherwise informal woodland setting. The combination of formal landscaping and informal plantings creates a garden of charm and distinction.

A historic tree collection provides the backbone of this 25ha garden which is set in the rolling landscape of Stobo Valley about 20km west of Peebles and 60km south of Edinburgh. At altitudes ranging from 165m to 250m, Dawyck is RBGE's best site for growing montane plants. In this climate, conifers and deciduous trees thrive and autumn colour is spectacular.

Three hundred years of tree planting

The 1968 hurricane devastated Dawyck Estate. With the destruction of 50,000 trees the estate lost 30 per cent of the standing value of its timber. The winds set in motion an extraordinary gift to the nation; despite the devastation, RBGE would fall heir to one of the most remarkable collections of historic trees in Scotland.

Dawyck had benefited from strong and unusually stable stewardship stretching back over 700 years. Only three families had owned the estate – the Veitches, Naesmyths and Balfours – and each one took a keen interest in trees and forestry. These pioneering planters left a lasting mark on the hillside where (winds permitting) trees live to a handsome old age. Veitches, lairds of Dawyck since the 13th century, were the first to plant horse chestnut (*Aesculus hippocastanum*) in Scotland and two of their original trees survived until 1932. At around the time James Sutherland was planting the Physic Garden in Edinburgh, the Veitches were planting the first European silver fir (*Abies alba*) in Scotland and one of them still stands in the Garden, more than 330 years later.

Sir James Naesmyth, who bought the estate from an indebted John Veitch in 1691, is best known for planting European larch (*Larix decidua*) at Dawyck in 1725 (not the first in Scotland as was once thought). His descendant, the fourth baronet, Sir John Naesmyth, was responsible for some of Dawyck's most magnificent conifers grown from seed collected during great expeditions ght   Like many other historic ardens in Scotland, Dawyck uffered terribly in the 1968 torm which tore down 50,000 rees throughout the estate nd caused structural damage o buildings. In the wake of this evastation the Balfour family ifted the estate's woodland arden to the nation in 1978 nd Dawyck became RBGE's hird Regional Garden. This mage, taken in 1982 from bove the Chapel, shows storm amage still very apparent.

ottom left   Fred Balfour lso had a passion for daffodils Narcissus). He planted swaths f them in and around the onifer and rhododendron lantings, as seen in this image rom 1934.

ottom right   The Balfour amily took over Dawyck Estate n 1897. F.R.S. (Fred) Balfour, een here beside a specimen f Rhododendron vernicosum n 1931, underplanted the tree anopy with many hundreds f rhododendrons and other hinese shrubs.

of the day, particularly those led by David Douglas. He added Scots pine (*Pinus sylvestris*) from seeds of the forests of Braemar and some of those still grow on the banks of Scrape Burn. He also found a curiously upright seedling beech which he transplanted close to the house. It would later be named *Fagus sylvatica* 'Dawyck' and has now grown to a height of 30m.

In 1897 the estate was bought by Mrs Balfour, the widow of Alexander Balfour whose company traded with Chile, Peru and the Pacific Northwest. Her son Frederick (generally known as F.R.S.) Balfour collected seed and plants for Dawyck on his travels to California. His greatest 'find' in 1907 was Brewer's spruce (*Picea breweriana*), and he arranged for 14 plants to be lifted and shipped to Dawyck in 1911, only the second introduction into Britain of this species. Forestry flourished at Dawyck as Balfour subscribed to the expeditions of George Forrest, Frank Kingdon Ward and, particularly, Ernest Wilson. Spring bulbs and rhododendrons added colour and Balfour benefited from contact with great collections such as the Arnold Arboretum in the US and RBGE. In 1914 Isaac Bayley Balfour sent a dozen

On Boxing Day 1998, Dawyck suffered a setback when a severe storm struck the south of Scotland and 34 trees were lost. Clearing storm damage put the planting programme on hold for two years. Despite this, the Garden was able to establish two interpretative trails in 1999. The David Douglas Trail (depicted left) leads visitors to species such as Douglas fir (*Pseudotsuga menziesii*), sitka spruce (*Picea sitchensis*), western hemlock (*Tsuga heterophylla*), grand fir (*Abies grandis*), noble fir (*Abies procera*), Lawson's cypress (*Chamaecyparis lawsoniana*) and western yellow pine (*Pinus ponderosa*).

The Rare Plant Trail was created as part of the Garden's Scottish Rare Plants Project to bring attention to the plight of some of Scotland's rarer plant species, such as wintergreen (*Pyrola minor*), oblong woodsia (*Woodsia ilvensis*) and sticky catchfly (*Lychnis viscaria*).

The David Douglas trail incorporates activities for children while the panels for the Rare Plant trail are placed at ground level close to the plants in question.

railway truckloads of rhododendrons to Dawyck.

Miraculously, most of this invaluable collection survived the 1968 hurricane. But the gales of 1973 compounded the damage, opening up land that was rapidly colonised by Balfour's ornamental but horribly invasive game cover, salmonberry (*Rubus spectabilis*). In 1978 Colonel A.N. Balfour (Frederick's son), recognising the value of his 300-year-old inheritance – and the monumental task that it would take to restore it to its previous splendour – decided to gift the arboretum to the nation. Dawyck became RBGE's third Regional Garden in 1979.

The canopy of the future

For RBGE a decade of hard work lay ahead: clearing storm debris, cutting back salmonberry, restoring paths and bridges, reopening vistas. It was at least five years before new tree planting was even contemplated and the first plantings of the mid 1980s were to strengthen the windbreak on the vulnerable

below left Dawyck's soil and climate are particularly suited to birches (*Betula*) and a fine collection stretches through the wild flower meadow from the Chapel to the Viewpoint. This stunning white-stemmed *Betula utilis* var. *jacquemontii* is grown from seed collected by Ron McBeath on 1 July 1985 at 3,000–3,500m on the Rohtang Pass, Himachal Pradesh, India.

below middle Staff planting a specimen of the Sicilian fir (*Abies nebrodensis*) as part of the International Conifer Conservation Programme. With fewer than 30 individuals remaining in the wild, it is probably the most threatened species of conifer in Europe and one of the most endangered in the world.

below right Imitating nature, species and cultivars of *Hosta*, *Rodgersia*, *Astilbe* and ferns have been planted in bold informal drifts, especially along the bank of the Scrape Burn.

western boundary. Dawyck Botanic Garden opened to the public in 1984; a treat for visitors discovering spectacular trees and stunning vistas; a triumph for the horticultural staff who had worked tirelessly to make the Garden safe and accessible.

At first the Garden was directed from Edinburgh's Arboriculture Department under Curator George Broadly with David Binns based at Dawyck as Supervisor. In 1992 David Knott – who had also worked at Logan – became Dawyck's first full-time Curator. His management plan was to build future development on the strong foundation of the past. The first task was to remove moribund trees to make space for new planting, not just to enrich the Living Collection but to add a multi-age canopy; a continuing insurance against storm damage. Virtually all the new

above Celebrating diversity: for more than ten years Dawyck has deliberately increased the variety of the Living Collection by adding more herbaceous plants, such as the Himalayan blue poppy (*Meconopsis* 'Slieve Donard') seen here against the spongy red bark of the giant redwood (*Sequoiadendron giganteum*).

Heron Wood Reserve

The Cryptogamic Sanctuary was established within the 2ha Heron Wood in 1996, following the success of the Cryptogamic Garden in Edinburgh. Heron Wood, a virtually untouched semi-native wood, had a canopy composed mainly of Scots pine, beech and self-sown birch. The Cryptogamic Sanctuary was established within the area of birch regeneration for the study and monitoring of fungi in particular, but also of mosses, lichens and liverworts. This area is not managed as intensively as the rest of the Garden – fallen trees, branches and leaf litter are left in place to be colonised by micro-organisms, which are abundant here thanks to the clean, undisturbed habitat. In terms of both its soil biodiversity and cryptogamic flora, the Sanctuary has been the subject of intense research and recording and is possibly the most intensively studied area of its type in the world. Interpretation panels explain the important contribution to our environment made by these tiny organisms and cryptogamic plants.

left Fruiting bodies of fungi such as the clouded agaric (*Clitocybe nebularis*) emerge through the leaf litter in the autumn.

material was of wild origin provenance and therefore of significant scientific value. It was carefully selected to complement the holdings at the other Gardens while taking advantage of the distinctive topography and microclimate at Dawyck.

Soon wild-sourced plantings were taking place in earnest, with new conifers, maple (*Acer*), *Berberis*, birch (*Betula*), *Spiraea* and rowan (*Sorbus*) all added. Bold herbaceous plantings – to diversify the collection and increase the Garden's appeal to visitors – also started in the early 1990s and continued for the next ten years. The Himalayan blue poppy (*Meconopsis*) brought vivid colour beneath North American conifers, while bold plantings of *Astilbe*, *Ligularia* and *Rodgersia* spread along the banks of the Scrape Burn.

In 1993 the staff at Dawyck started experimenting with grass cutting regimes, varying the timing and height of the cut, to encourage local wildflowers to become established within the sward. As a result, the common spotted orchid (*Dactylorhiza fuchsii*) appeared for the first time in 1999, and its numbers have increased each year since. Cut grass paths through wildflower meadows are now a feature of Dawyck's landscape. Nature has also been given the upper hand in the ancient Heron Wood

where, beneath old beech and young birch, the Cryptogamic Sanctuary has become the focus of international research into fungi.

Conifers have always thrived at Dawyck and so, from 1995, the Garden started adding further species. A grove of dawn redwood (*Metasequoia glyptostroboides*) was planted as part of the International Conifer Conservation Programme (ICCP). Further ICCP plantings followed, especially of fir (*Abies*), juniper (*Juniperus*), larch (*Larix*), spruce (*Picea*), pine (*Pinus*) and hemlock

opposite *Rhododendron calophytum* growing over the Scrape Burn much as it might have done in its native habitat in Sichuan, China. This specimen predates RBGE's management Dawyck and no records remain but it is thought to be a Wilson introduction.

right Dawyck's Curator, Graham Stewart, collecting seeds of sweet gum (*Liquidambar styraciflua*) near the Hiwassee River, Polk County, Tennessee in 2010.

far right Past and present Regius Keepers gather with staff and friends at Dawyck to unveil a sculpture dedicated to Douglas Henderson (Regius Keeper from 1970 to 1987), who negotiated the transfer of Dawyck to RBGE. Left to right they are: David Ingram (Regius Keeper 1990–1998), Stephen Blackmore (1999 to the present) and John McNeill (1987–1989).

(*Tsuga*). In 2006 wild origin Sicilian fir (*Abies nebrodensis*) was established high on the hillside, offering a safe haven to one of the world's most endangered conifers.

Even historic trees do not last for ever. The Beech Walk was a well-known avenue of stately beech trees lining a grass terrace which had fine views of Dawyck House and the hills beyond. By the mid 1980s, however, many of the trees had become over-mature with massive branches overhanging the Walk. For public safety it became necessary to remove many of the limbs and, over the years, whole trees have also had to be removed. In 2010 semi-mature replacement trees were planted. In years to come this landscape feature may once again make a bold statement.

In 2008 David Knott became Curator of

opposite A gnarled old European larch (*Larix decidua*), planted in Dawyck in 1725. This stooped and spindly specimen, with snapped and cracked branches, has survived hurricanes and hard frosts for almost 300 years and shows no sign of giving up yet.

right top Staff cutting up a felled beech tree (*Fagus sylvatica*).

right bottom Staff clearing up after arboricultural work.

the Outdoor Living Collection at Edinburgh and Graham Stewart took over as Dawyck's Curator. Graham was first attracted to Dawyck by the trees in 1986 and he can measure his time here in the healthy growth of a young sequoia outside his office window. New trees and flowering shrubs continue to arrive and thrive in the Garden, grown from wild origin seed collected in cooler montane regions across the world. The canopy of the future grows in the shelter of trees planted hundreds of years ago. They are rightly regarded as our heritage. In 2002 some of Dawyck's oldest and finest specimens received Heritage Tree Awards. Included in a selection of one hundred Heritage Trees of Scotland are a European larch (*Larix decidua*), planted by James Naesmyth in 1725, and a European silver fir (*Abies alba*), planted by John Veitch in 1680.

Dawyck has come a very long way since the devastating storms of 1968 and 1973. The Garden now looks magnificent and in 2007 a new environmentally aware visitor centre was erected along with new storage facilities and workshops. In 2009 Dawyck was awarded five star visitor attraction status, the first garden in Scotland to be so awarded and this was followed in 2010 by a Silver Green Tourism award.

113

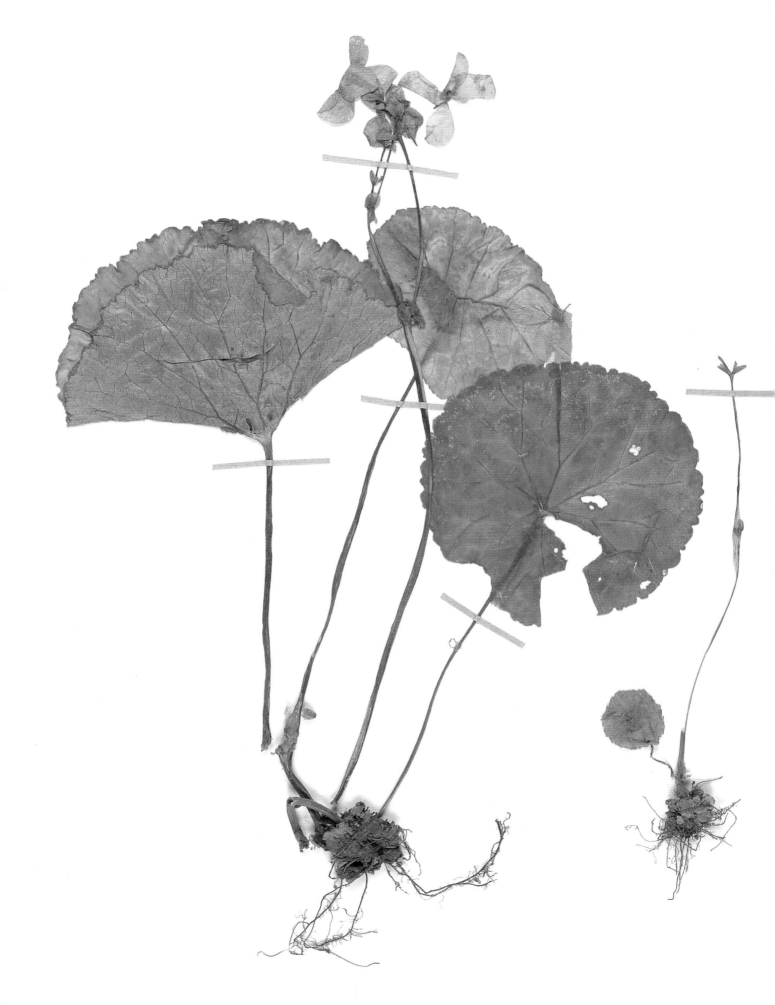

2
Curating the Living Collection

Collecting plants

EXPEDITIONS AND PLANT COLLECTION

The Garden is always growing. If RBGE is to maintain its global status as a leading botanic garden and to increase its role in global plant conservation, it must constantly develop its Living Collection by bringing in new plants from expeditions around the world, exchanging plants with other botanic gardens and adjusting to meet the ever-changing needs of research and conservation.

left Staff and Chilean colleagues collecting seed, herbarium specimens and leaf samples for DNA analysis from plants in northern Chile for research in Edinburgh and Chile and to strengthen the Living Collection at Edinburgh.

The world's botanic gardens each have their own missions and specialisms and therefore concentrate on growing the plants needed for their own work. No garden would ever attempt to grow all vascular plant species, estimated at in excess of 380,000 worldwide, or the countless thousands of cultivars, and so most botanic gardens have a collection policy to ensure that the right plants are available for those who need them. Some policies simply list the plants that the garden aspires to cultivate, while others are more detailed and include strategies, guiding principles and management procedures.

RBGE's *Collection Policy* guides the development of the Living Collection to ensure that the appropriate plant species or geographic regions are prioritised and to establish comprehensive procedures and targets. The current policy, published in 2006, identifies the users of RBGE's Living Collection and states the Garden's aspirations and methodologies. It ensures that development of the Living Collection complies with the Garden's wider agenda – including the research, conservation and education work as well as the Garden's international context and commitments.

The *Collection Policy* governs all aspects of the Living Collection, aiming to create the richest plant collection possible within the resources available. It helps to provide continuity and to guard against short-term

policy changes that could impact the Collection adversely in the long term. For example, long-term planning is particularly important for trees – it is essential to plan many years ahead to create a multi-age and variable density canopy in order to to provide shade and wind shelter for other plants and to plan for the trees of the future.

At the heart of the *Policy* are guidelines for acquisition – setting priorities for new plants coming into the collection. This includes guidance for fieldwork such as issues of legislation and standards of

ROYAL BOTANIC GARDEN EDINBURGH
Collection Policy for the **Living Collection**

David Rae (Editor), Peter Baxter, David Knott, David Mitchell, David Paterson and Barry Unwin

recording information in the field. RBGE encourages close collaboration between its horticulture and science staff as well as partnerships with other institutions, especially those in the host country.

The plants that come into the Living Collection are required to be well documented and collected in the wild wherever possible. Species are categorised by priority for acquisition, with the highest priority given to those in which RBGE has a major research interest. The *Collection Policy* ensures that the most appropriate plants are cultivated at each of RBGE's four Gardens, making use of the different climates, topographies and soils. It also gives guidance on representation – how the plants are actually laid out in the Gardens. This combines the taxonomic approach, where all the species within a genus or family are grouped together, with the geographic approach, where plants are grouped by geographical origin or by habitat.

above Planting fashions change over time. Chapter 4 of the *Collection Policy* describes different approaches to laying out plants in the Collection. Two of the most enduring techniques are to adopt a geographic approach seen at Benmore in the Chilean Hillside (top left) or a taxonomic approach where species of the same genus are grouped together for easy comparison such as the Birch Lawn at Edinburgh (top right).

The fact that our climate is now known to be changing faster than at any time in history will inevitably have consequences for our plants, and so the *Collection Policy* draws attention to the potential impact of increasingly erratic weather patterns. Staff are advised to predict groups of plants at all four Gardens that may need to be relocated, to take note of any new sightings of pests or diseases and to be alert to opportunities to grow new plants, exploiting the four Gardens to the full and developing innovative conservation programmes.

More plant species in the wild are threatened than ever before and so plant conservation is integral to the thinking throughout the *Policy*. It calls for special management for conservation collections – groups of plants held at RBGE as an insurance against loss or genetic erosion of the species in the wild. Threatened species intended for recovery and restoration programmes are assigned to each Garden for cultivation. The importance of RBGE having rich collections of British, and especially Scottish, native plants is also emphasised for the Garden's national status, for teaching, research and conservation, as well as for interpreting our natural history.

Ensuring that the Living Collection is fit for purpose involves making the Gardens as accessible and attractive as possible, thereby creating beautiful landscapes for recreation, contributing to visitors' quality of life and increasing visitor numbers. The Gardens are also of great historic value. The Edinburgh Garden is more than 180 years old (on its current site at Inverleith, although it was founded in 1670) and the Gardens at Logan and Dawyck were established long before RBGE acquired them. Thus RBGE has numerous champion trees, plants associated with famous collectors, early introductions and important landscape features such as historic designs and listed buildings. The *Policy* serves as a reminder of our

right Communicating research and conservation of Scottish plants is an important part of RBGE's work and the *Collection Policy* lays out guidelines for the cultivation and display of Scottish native species in each of the four Gardens. Here at Logan (right) sticky catchfly (*Lychnis viscaria*), thrift (*Armeria maritima*) and globe flower (*Trollius europaeus*) thrive in a recreation of a seaside habitat.

International conventions

The Collection exists within a local, national and global context. It supports the UK and other countries in fulfilling their national and international obligations, with specific reference to the following:

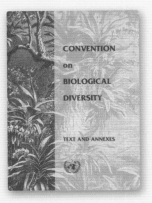

Convention on Biological Diversity

The purpose of the Convention on Biological Diversity is threefold: the conservation of biological diversity, the sustainable use of biodiversity and the fair and equitable sharing of the benefits arising from biodiversity. It gives RBGE a strong mandate to work on conservation, while specific articles govern access to genetic resources, covering issues such as permission to collect, access agreements and the transfer of collected plant material to third parties.

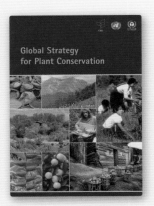

Global Strategy for Plant Conservation

The Global Strategy for Plant Conservation (GSPC) is probably the most relevant current agenda for plant conservation as far as botanic gardens are concerned. The Garden is contributing to at least eight of its sixteen targets, including those

covering integrated conservation programmes, the cultivation of threatened native plants, the dangers of invasive species, education, networking and sharing skills, and expertise through capacity building in other countries.

Plant Diversity Challenge

Plant Diversity Challenge is the UK's response to the GSPC and reports on the progress made in each target following detailed consultation with all involved. After an explanation of what each target means within the UK context, the report details relevant ongoing work and prioritises necessary additional actions which will enable the UK to meet the targets. These include matters such as assessing priorities, developing methodologies for conserving plants outside their natural habitat and maintaining the genetic diversity.

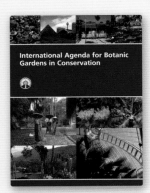

International Agenda for Botanic Gardens in Conservation

This provides a global framework for the development of botanic garden policies and programmes for the effective implementation of international treaties and national

laws and strategies relevant to plant conservation. The *Agenda* defines the role of botanic gardens in the development of global partnerships and alliances for biodiversity conservation, and presents a means to monitor this work. Botanic gardens can contribute by developing the quality of their plant collections, maintaining genetically diverse collections and developing best practice in conservation programmes.

Action Plan for Botanic Gardens in the European Union

Published in 2000 in response to the need for coordination and unity between botanic gardens, the *Action Plan* sets out more than thirty objectives grouped in six major themes ranging from horticulture and science to heritage culture and tourism. RBGE's *Collection Policy* endorses the *Action Plan* and encourages the use of the plant collections to support its objectives.

European Plant Conservation Strategy

Planta Europa is a network of organisations working to conserve wild plants and their habitats

in Europe. *The European Plant Conservation Strategy* (EPCS) has set major targets, many of which echo the GSPC, such as understanding and documenting plant diversity, using plant diversity sustainably and promoting educational awareness. RBGE supports the EPCS by providing listings of all threatened European plants in cultivation as well as case studies and best practice guidelines for integrated plant conservation programmes.

Convention on International Trade in Endangered Species of Wild Fauna and Flora

The purpose of Convention on International Trade in Endangered Species of Wild Fauna and Flora (CITES) is to regulate and monitor trade in endangered species (living and dead) thereby reducing the market for, and financial reward gained from, plundering plants – and animals – directly from the wild. RBGE does not trade in plants commercially, but in the context of CITES, 'trade' is taken to mean the crossing of a national border. On occasions, RBGE does collect CITES-registered plants from the wild and its policy is to abide by CITES regulations at all times (for instance, by obtaining licences to collect CITES-listed species) and to cooperate with customs staff in identifying plants and holding plants confiscated from others because of infringements of CITES.

responsibilities for the historic elements of our plant collections and landscapes, and of their cultural and scenic value.

As the *Collection Policy* covers every aspect of curating the Living Collection, RBGE hopes that it will be relevant and comprehensive to all who want to use it. Above all, it aims to ensure cohesive and careful stewardship of our plant collections, raising the standards of acquisition, cultivation and record keeping, and enhancing the Collection's multi-faceted value: for science, conservation, education, heritage and beauty.

Collecting in the wild

New plants come into the Garden on an almost daily basis. They are brought back from expeditions around the world or acquired through exchanges with other botanic gardens, *indices semina* (botanic garden seed lists) and purchases from nurseries. Of these possible sources, field-work is by far the most important because it is only by working in the country where the plant grows in a wild state that seeds can be collected, herbarium specimens

obtained and field data recorded. It is this so-called 'wild origin' material that is of great importance to scientists and conservationists when studying a plant from the Living Collection, as they need to be sure that it is a true representative of a population from a particular place in the wild. The Collection might, for instance, contain ten accessions of a species. If they had been bought from a nursery then little could be derived from them except their appearance. If, however, each had been collected from a different population in the wild along with a herbarium specimen and detailed field data, then each may be regarded as a representative of that population. So, if well selected and recorded, ten plants growing together in the Garden could, in fact, represent the diversity of the entire species, perhaps spanning several thousand square kilometres.

As well as the significant benefit of obtaining these plants for research and conservation work, fieldwork offers the added value of staff seeing plants growing wild, which invariably leads to better cultivation back at the Garden.

RBGE's active research programmes in

above Plants can be bought from nurseries or exchanged with other gardens, but only collecting plants in their natural habitat provides a full set of data. Wild origin (or wild source) material is essential for research and conservation and is the preferred source of incoming plant material at RBGE. Each plant in the Collection can be traced back to a specific place at a specific time and data such as altitude, aspect and associated vegetation can be recorded.

top left The Terai region of Nepal, 2004.

bottom left Gaoligong Shan, south-west China, 2002.

right New Caledonia, 2007.

CITES

CITES is possibly the most powerful of the international biodiversity conservation agreements because its provisions are translated into national laws in all the countries that sign up to it. The Convention has been in force for over 30 years and has been ratified by over 170 countries. Botanic gardens have always played an important role in the implementation of CITES for plant species and in raising awareness of the aims and requirements of the Convention. Much of the botanical expertise and information needed to ensure the effective operation of CITES continues to be provided by staff in botanic gardens.

CITES provides a mechanism to regulate and monitor the international trade in threatened wild plants or plants which may become threatened in the future without regulation. It does this through categorising plant species into three levels of protection from over-exploitation. The three Appendices of the Convention relate to these three levels. Appendix I lists the most endangered species, those which are threatened with extinction, and includes approximately 300 plant species. Appendix II lists species that

are not necessarily now threatened with extinction but that may become so unless trade is closely controlled, including 28,000 plant species and the entire orchid and cactus families (such as this species of *Copiapoa*, from northern Chile). Appendix III is a list of species included at the request of a Party that already regulates trade in the species and that needs the cooperation of other countries to prevent unsustainable or illegal exploitation. International trade in specimens of species listed in this Appendix is allowed only on presentation of the appropriate permits or certificates. Only a handful of plants are listed in Appendix III but over the last 15 years some countries have used this Appendix to help control international trade in certain tree species. International trade in any of the listed species is prohibited but botanic gardens (and others) may apply for a licence to collect small numbers for research purposes only.

On occasions RBGE needs to collect CITES-listed plants from the wild for research. The Garden's policy is to ensure relevant staff abide by CITES regulations at all times and keep any documentation up to date in computerised databases.

many different countries entail a substantial amount of paperwork to ensure the Garden complies with various laws, conventions and plant health regulations. The Convention on Biological Diversity (CBD), the Convention on International Trade in Endangered Species of Wild Fauna and Flora (CITES) and a number of plant health regulations are just some of the important areas of national and international law that staff need to be aware of, whether engaged in fieldwork or back at the Garden.

Fieldwork

The underlying purpose of fieldwork is always to obtain data for future analysis, research or conservation and to augment RBGE's living and preserved plant collections. An expedition can be initiated in many different ways: sometimes staff are

invited to visit a country because of their known expertise in a region or plant group. On other occasions, the work is initiated at RBGE to enhance an existing programme of work. For example, the purpose may be to collect data from less visited regions within a country of particular interest, as was the case in RBGE's 2004 expedition to the Terai region of southern Nepal. Many of Nepal's northern alpine locations had already been sampled for herbarium specimens and living material but, with the launch of the *Flora of Nepal* project in 2002, staff discovered that there were few specimens from the lower elevation, southern regions of Nepal and this hampered progress on writing accounts for the *Flora*. On other expeditions, large amounts of data may be required from a fairly small group of plants within a specific geographic area. This was the case in the four trips to New Caledonia over the past ten years, when numerous samples of leaf material were taken from species of *Araucaria* to investigate the taxonomy and population boundaries of the various species found in the region.

Permissions

The first, and sometimes most difficult, step in planning an expedition is to get permission from the country concerned to work in the field and to take herbarium and seed samples. All countries have sovereign rights over their biodiversity and so third parties need permission to collect material. This applies even if the material is not going to be used commercially – the Garden only seeks to take specimens for research, conservation, education and display. Even so, gaining such permission can be a protracted process and each country has different regulations and procedures. Obtaining permission to collect plant material from another country

can take a considerable amount of time and correspondence and so staff are encouraged to start the process up to two years before the planned date of departure. The best route is to build up local and government contacts over many years so that staff get to know the people involved and how the system works.

It may also be necessary to obtain phytosanitary documentation to bring plants into the country. A phytosanitary certificate is a statement issued by the plant health authorities in the exporting country that the material to which it relates has been officially inspected in the country of origin or country of despatch, complies with legal requirements for entry into the EU and is free from certain serious pests and diseases. For some threatened species such as orchids, cacti and succulents, CITES permits may also be required both to take these plants out of a country and to import them into another country.

Temperate expeditions – Japan

The Garden has a long-term interest in plants from the Sino-Himalayan region including the Himalaya, of which Japan is located on the eastern extremity. Before this century, however, although the Living Collection contained a number of species from Japan, few had been collected in the wild and records were poor. The Garden therefore had rather weak holdings of species from this floristically rich and important country. To enhance the Collection, an introductory visit to the north of Japan was made in 2003 (EJE, see right) with the intention of following this up with a further visit in 2006 or 2007 to the south. Fortuitously, staff from the Garden were invited to join two other expeditions (BBJMT in 2005 and BCJMM in 2007) and these, along with the second Garden-initiated visit in 2006 (ESJE), now bring the total number of visits to four. In total 1,334 living accessions and 1,177 herbarium specimens have been collected on these expeditions, greatly increasing the Garden's holdings of Japanese plants. While many individual plants from these expeditions are now growing in Edinburgh,

Logan and Dawyck, the majority are being planted at Benmore in an area dedicated to Japanese plants.

EJE 2003 (Edinburgh Japan Expedition) Kenji Kano, David Knott and David Rae Northern Honshu and Hokkaido 306 accessions

BBJMT 2005 Tim Bolton, Peter Brownless, Robert Jamieson, Bill McNamara and Shigeto Tsukie Northern and Central Honshu 312 accessions

ESJE 2006 (Royal Botanic Garden Edinburgh Southern Japan Expedition) Peter Baxter, David Knott and Matsushita Hirotaka Southern Japan 360 accessions

BCJMM 2007 Peter Brownless, Tom Clark, Robert Jamieson, Vince Marrocco, Bill McNamara and Shigeto Tsukie. Northern and Central Honshu 312 accessions

top Hakusan National Park in the Chūbu Region of Honshu, Japan, 2003.

bottom left Team members wading through bamboo in southern Japan, 2005.

bottom right Peter Baxter, Curator of Benmore Botanic Garden, and David Knott, Curator at Edinburgh, collecting seeds on the slopes of Mt Aso, which is the largest active volcano in Japan, in the Kumamoto Prefecture, 2006.

CBD

above CBD logo: the Convention on Biological Diversity (CBD) has three objectives (explained on the right). Since its adoption in 1992 permission to collect plants in other countries has become mandatory rather than common courtesy as it was before – any plant material found within a country is the property of that country. RBGE requires its staff to obtain written permission to collect plants in other countries prior to departure.

The Convention on Biological Diversity

The process of acquiring plants has changed radically in recent years because of the Convention on Biological Diversity (CBD), which was signed by 157 countries following the Earth Summit at Rio de Janeiro in 1992.

The purpose of the CBD is threefold: the conservation of biological diversity; the sustainable use of its components; and the fair and equitable sharing of the benefits arising out of the utilisation of genetic resources. In the context of taking plant specimens out of another country, the third objective is the most relevant. Prior to the adoption of the CBD, all living organisms in any country were considered to be freely available to anyone, although it was always considered common courtesy to try to obtain some official permission before collecting plants. At the Earth Summit this issue was debated in detail and it was eventually decided that the biodiversity residing in a country was the property of

Standard information to be recorded when collecting plants in the field

Source: The name and address of the institute or individual that donated the material (this may be the same as the collector)

Field name: The 'working name' of the plant material. This may be quite vague and only to a genus or family level due to time constraints or if the plant is not in flower

Type of material: i.e. seed, cutting, rhizome, leaf tissue, DNA etc

Provenance: An indication of the origin of the material using codes from BG-BASE, the plant records system used at RBGE, for example:

  W: collected directly from the wild
  Z: from a cultivated plant of known wild origin
  G: from a cultivated plant not of known wild origin

Genetic variability: The sampling method used, for example:

  XX: Sample method unknown

SO: Seed from one plant
SM: Seed from more than one plant
VO: Vegetative material from one plant
VM: Vegetative material from more than one plant

Collector or expedition code: Each different collecting group or expedition is identified by a unique code. These are usually composed of three, four or five letters and/or digits, for example:

  ACE: Alpine Garden Society Expedition to China, 1994
  ICE: Instituto de Investigaciones Ecológicas Chiloé (IIECH) & RBGE
  ESIK: RBGE Sikkim Expedition, 1992
  GSE96: Gaoligong Shan Expedition, 1996
  GSE97: Gaoligong Shan Expedition, 1997

Collector number: Identifies a particular collection made during an expedition or collecting trip.

Together with the collector/expedition name/code they provide a unique combination that identifies the collection whether it be herbarium specimens, living material, DNA samples etc, wherever it is placed. Collector numbers are simply allocated in chronological order and are quoted along with the collector/expedition name, for example ACE 1, ACE 2, ACE 3 etc.

Date of collection

Country of collection: Along with administrative divisions if possible, for example:

Datisca cannabina: Turkey, Ankara Province or *Libertia sessiliflora*: Chile, Region V, Prov. de Petorca, Zapallar

Locality and altitude: A descriptive account with latitude and longitude where possible (preferably from GPS), for example:

Libertia sessiliflora: Chile, Region V, Prov. de Petorca, Zapallar; Quebrada Aguas Claro, 2km S of Cachagua;

3km inland; 32°35's, 71°55'w.

Habitat or environment from which the material was collected, for example:

Cistanthe grandiflora: crevices of rocky cliffs above sparsely vegetated shingle beach.

Associated plants: A listing of other species growing nearby, for example:

Sorbus arranensis: Associated ground flora including *Vaccinium myrtillus*, *V. vitis-idaea*, *Sphagnum capillifolium*, *Blechnum spicant*, *Populus tremula* and *Sorbus aucuparia*.

Description of the material emphasising characters that may not be evident from the material after collection (such as colour), for example:

Calliandrina grandiflora: succulent-leaved herb to 50cm; flowers nodding, petals cerise-pink, sepals mottled.

that country and that anyone wishing to access it or use it in any way would need to get advance permission to do so from the country in question. For botanic gardens wishing to collect plant material overseas, this clearly had considerable implications in terms of paperwork and permits.

RBGE strives to set the highest standards of compliance within the spirit and law of the CBD and in recognising its responsibilities has signed two important agreements that lay down protocols for collecting and transferring plant material. The first is 'Principles on Access to Genetic Resources and Benefit-sharing', which is concerned with how to obtain official permission to collect plant material directly from the wild in another country. The second is the International Plant Exchange Network (IPEN), which concerns the exchange of plant material between botanic gardens, based on the prerequisite that the two gardens concerned sign up to certain standards of regulation compliance and that the plants to be exchanged were collected legally in the first place.

Planning

Time spent in the field is expensive and so it pays to plan everything meticulously to ensure that the maximum benefit is made from the investment. Fieldwork requires careful planning to make sure that provision is made for access to an area, travel, food and shelter and, of course, the necessary botanical work to be done.

below Fieldwork expeditions require meticulous planning to ensure that time in the field is well spent and that the objectives of the trip are achieved. Here, staff are discussing fieldwork in northern Chile, 2008.

Modes of transport: on two feet or four, by air or over land: RBGE staff use almost every mode of transport conceivable to obtain plants and essential data for the Living Collection and the research and conservation it supports.

top left Helicopter travel is expensive but it has the advantage of placing expedition members and their equipment deep into remote areas quickly, avoiding the time required to walk into the site. Nepal, 2008.

bottom left 4 x 4 vehicles are the only realistic way of carrying enough food, water and equipment needed in large desert regions. Atacama Desert, Chile, 2008.

Fundraising is an important part of expedition planning as fieldwork in an overseas country can be very expensive. As well as having to cover the cost of travel to and within the country, it is customary to pay the travel expenses of local colleagues. Porters may need to be hired and four wheel drive vehicles may be necessary. Add to this hotel costs at the start and end of the visit, equipment purchase or hire, food and medicines and, possibly, excess baggage costs to transport material back to Edinburgh, and the sums involved can become quite substantial. RBGE is usually able to provide some funds to support the basic costs of fieldwork but all expeditions have to seek additional income from trusts, grant-giving bodies or commercial sponsorship.

Plans also have to be made for transport in the field. This may involve using vehicles or horses, or porters to help carry the equipment if trekking. Routes need to be planned, equipment assembled and provisions sought. Once in the field it is important to be able to get on with the job of collecting plants and recording information, and so any time spent repairing vehicles, mending broken kit and sourcing supplies is time wasted. Itineraries need to be planned carefully and a combination of experience and local knowledge is invaluable in judging the time needed to cover an area of ground and to complete the necessary work.

Characteristics of an expedition

The terrain, the climate and the objectives of the fieldwork decide the itinerary: an expedition to the Himalaya will be dictated by landform, passes, altitude and accessible valleys, whereas in desert regions it will be decided by sources of water, firmness of sand and the physical restrictions of the terrain.

top right Trekking is the only viable means of getting around in the mountainous regions of south-west China but crossing rivers such as the River Dabadi on the 2002 Gaoligong Shan Expedition can be a nerve-wracking experience.

bottom right In the Himalayas pack horses (or sometimes yaks) are often used for carrying food and equipment although porters are more frequently employed. Nepal, 2008.

Where the object is to study a very specific group of plants, for instance to harvest DNA samples from several individuals within a single genus, it may be possible to stay in a local hotel. For example, in the four expeditions to New Caledonia, the team has been able to drive out to the plants in question each day to obtain the material and record the associated information before returning to the hotel in the evening to write up the notes and package the collections.

Other types of fieldwork necessitate covering large areas of ground to sample a whole range of species from a particular region. In mountainous countries such as China and Nepal, this is most easily achieved by trekking. Such expeditions may take four to eight weeks and involve numerous porters.

For example, in 2006 RBGE staff took part in an expedition to the Gaoligong Shan, a range of mountains in south-west China running along the Yunnan Burma border. The group trekked to Mt Gawa Gapu (the so-called 'Path of Terror' – a description coined by the plant collector Frank Kingdon Ward because of the dangerous nature of part of the terrain) and the fieldwork lasted for six weeks and involved over forty porters. This trip was part of a series which took place over five years, organised primarily by the California Academy of Sciences with the purpose of cataloguing the entire biodiversity – plants, animals and insects – of the Gaoligong Shan.

In flatter terrain, such as desert or semi-desert regions, where long treks are not possible, a four wheel drive vehicle is the best solution. This was the transport used in the 2008 expedition to the Atacama Desert in northern Chile to collect distribution data about threatened endemic species and also to collect seeds for Edinburgh's Living

above Riding high: collecting by elephant makes it easier to collect specimens from trees and saves the effort of trekking.

Tropical expeditions – Sulawesi

RBGE has a history of plant research in South-East Asia dating from 1962, when two staff undertook fieldwork in Sarawak. In January 2000, a joint field trip (ARMEH) was undertaken to western Sulawesi by RBGE and Kebun Raya Bogor, one of four botanic gardens comprising Indonesia's National Botanic Garden (KRIB), located in West Java. This was the first time RBGE had visited the island and it proved to be very successful. In March 2002 another expedition visited botanically unexplored areas of Gorontalo province, in northern Sulawesi. The 2004 expedition to Central Sulawesi (Sulawesi Tengah) was a further extension of this collaborative research.

The aim of the expedition was to collect herbarium, spirit, DNA and living material of families of research interest to RBGE and KRIB, particularly Gesneriaceae, Zingiberaceae and Ericaceae. The team as a whole also undertook general herbarium collecting, in multiple sets, of material of interest to KRIP, RBGE and other institutions working on the *Flora Malesiana* project. Coordinated by Leiden University (Netherlands), the aim of this project is to define, name and provide identification guides to the plant species of the Malesian region. It involves research institutions in many countries and is documenting the flora family by family, providing the primary taxonomic and biodiversity information that is basic to any conservation measures.

The living material collected (Gesneriaceae, especially *Aeschynanthus*, *Agalmyla* and *Cyrtandra*, Zingiberaceae,

especially *Alpinia*, *Amomum* and *Etlingera*, *Rhododendron* of section *Vireya*, and certain Asclepiadaceae) was to be grown at KRI-Cibodas, Bogor (another of the four botanic gardens, also in West Java) and at RBGE, and used for research, public education and display, and for *ex situ* conservation.

Collecting was carried out on two mountains, Gunung Katopas and Gunung Hek. A total of 319 herbarium collections, 221 living collections and associated leaf samples in silica gel were made, representing 61 families and 70 genera. The specimens included a number of new species and several new records for the island, enabling meaningful research to continue and contributing to important biodiversity documentation, necessary for developing conservation strategies. The existing link with Indonesia's National Botanic Garden and National Herbarium was further strengthened and a new link with Herbarium Celebense, Palu was made.

ARMEH, 2000 George Argent, Mary Mendum, Louise Galloway, Paul Smith and Hendrian, western Sulawesi 373 accessions.

SUL02, 2002 Mark Newman, Steve Scott, Mary Mendum, Hannah Atkins and Hendrian, northern Sulawesi 205 accessions.

HNSS, 2004 Mark Newman, Steve Scott, Nazare Saleh, Hendrian and Dadi Supriadi, central Sulawesi 221 accessions.

above Mark Newman, Hannah Atkins and Mary Mendum pressing herbarium specimens in Sulawesi, 2002.

Collection. In these situations it is possible not to have access to a water or fuel source for a week or more unless it is taken on expedition.

Other expeditions, such as the 2005 fieldwork expedition to Tasmania, New Zealand and Australia, involve varied terrain. Here the team combined trekking and driving – the best solution for exploring both the dense forests and mountainous terrain. The purpose of this trip was to strengthen the Living Collection at Logan – 560 accessions together with herbarium specimens were collected, including species of *Eucalyptus*, *Callistemon*, *Leptospermum* and ferns.

Over many years of exploration around the world, Garden staff have used a range of transport including horses and elephants as well as the usual boats, buses, trains, cars and planes. Even when on foot, the terrain, weather and vegetation can present diverse challenges. RBGE's scientists have experienced steep climbs in the mountains, waded through bogs, rivers and lagoons and trekked through dense jungle using machetes to create new tracks. Expedition staff have had to face relentless monsoon rain, deep snow, intense heat and dust storms, not to mention mosquitoes, snakes and even encounters with unpredictable yaks in Nepal and Bhutan.

Collecting seeds

The objectives of the fieldwork will govern what is collected and how it is handled and stored. Wherever possible, herbarium specimens are collected alongside every seed collection; this is best practice even though the plant might not be in flower. An exception to this might be when seed is collected from a herbaceous plant late in the autumn when there is no herbarium material worth collecting.

Seeds come in all shapes and sizes and are enclosed in a similarly vast array of 'packages', from tiny, dry, dusty capsules through to big squashy fruits. These all need to be processed and cleaned during fieldwork to end up with insect- and debris-free seeds stored in labelled packets by the end of the trip, ready for sowing once back at Edinburgh.

Some seeds and dry, indehiscent fruits (which have seeds that are not enclosed in a capsule) can be placed directly into paper packets in the field for later processing. However, the majority of collections are usually placed into 'bird bags'. These are small cotton bags with a drawstring at the top. Bird bags are useful for drying dehiscent fruits

top left Extracting seeds from the dried seedheads of *Rhodophiala bagnoldii* before separating the seeds from the debris and then packaging and labelling. Northern Chile, 2008.

top right Cotton bags with drawstrings are perfect for drying seeds which have been extracted from juicy fruits. Dried seeds are decanted into paper seed packets. Here, a member of staff checks the condition of seeds. Nepal, 2001.

bottom left Cotton seed bags, commonly referred to as 'bird bags' – they are also used by ornithologists to hold birds for ringing or research – hang on ropes slung between tents to dry. Makalu region of Nepal, 1991.

bottom centre Much time and energy is devoted to collecting and cleaning seeds and care is taken during transportation to ensure they are kept safe and at the right temperature. Here on the Edinburgh Makalu Expedition of 1991 the job of transporting pressed specimens and seeds was given to the most trusted and sure-footed Sherpa.

bottom right Fleshy fruits are squeezed and squashed to extract seeds. Once fruits are squashed and partially dried, seeds can be extracted from the pulp before cleaning, further drying and packaging. Processing *Sorbus* seeds in Japan, 2003.

(which have the seeds enclosed in a capsule) as the fruits can be kept until they split and the seed is released. These bags are essential for fleshy fruits that need to be cleaned and dried before they go mouldy.

The 'pulp', or fleshy part of the fruit, in many genera contains an inhibitor which needs to be removed to allow germination to take place. Depulping also allows the individual seeds to be handled more easily. Fleshy fruits (*Sorbus* and *Arisaema*, for example) are placed in the bag, the top knotted and the whole bag crushed to split the fruit. This can be spectacularly messy for juicy fruits and the excess liquid needs to be squeezed out before it has any chance of drying. Another way of separating the pulp from the seed is to immerse the squashed seeds into a bowl of water as the pulp falls to the bottom of the bowl whilst the viable seed floats to the top. The seed bags can then be air dried on lines or spread out over warm rocks in the sun, or can be suspended on strings over a heated drying frame. Some plants with fleshy fruits, such as many species of myrtle (*Myrtus*), fail to germinate if the seeds are dried out too much and must therefore be kept moist.

For dry seed, particularly of genera such as *Primula* and *Meconopsis*, which are tiny and can be difficult to separate from the broken-up capsule, a variety of fine sieves of different sizes can be used to good effect. The seeds are generally of similar size to each other and smaller than the capsule debris and so can be separated relatively easy.

Once fleshy fruits are dry, or dry dehiscent fruits have shed their seed, the seed can be removed from the bag, cleaned and put into packets. The finished seed packets are sealed

Recording essential data: collected plants and seeds are only as valuable for future research as the accompanying field data – altitude, aspect, plant size and habitat. This information is transferred onto the Garden's database on returning to Edinburgh.

top left Some species make poor herbarium specimens when their floral parts are thin and delicate or it is difficult to appreciate their three dimensional form once pressed. Here, staff from Edinburgh and a colleague from Iran have undertaken a floral dissection of *Iris acutiloba* ssp. *lineolata* and are photographing the individual parts against a black velvet cloth. Iran, 2005.

top right From field to database: large field notebooks have been used for generations to record the field data listed on p. 126; now (as seen here in Oman in 2003) laptop computers are used more frequently – data can be downloaded directly into the Garden's plant records database on return.

with tape to stop small seeds escaping, and the packets stored in a cool, dry place. The best way to store and transport them is to place them into plastic boxes with holes punched in the tops to allow excess moisture to escape. They should, however, be inspected regularly to ensure that they are not going mouldy.

If carried out properly and patiently, the cleaning of seed is a time-consuming and laborious task but it is a vital part of plant collecting fieldwork and is essential for successful germination. Once an expedition comes to an end it is important to transport all the seed, herbarium specimens, living material, data and equipment back home. It can take a considerable amount of time to prepare the material for transport home and, where necessary, obtain final documentation. It is important to package the material carefully for safe transport, to avoid wasting all the effort devoted to obtaining it in the first place.

Data recording in the field

The data attached to each plant or accession makes it potentially more valuable for research or conservation than the plant

alone, and so botanic garden staff spend considerable time in the field not only collecting, cleaning and storing seed, but also recording field data and entering the information in collecting books and, more frequently nowadays, into laptop computers.

Detailed and accurate plant records are essential to the efficient management of living collections and for their effective use in conservation, education and research. Provenance details, including information about the plant's natural habitat and surrounding vegetation, add enormous value to a specimen and can never be added later if lost or not recorded in the first place.

The Garden deals with increasingly complex plant records issues and so produced a handbook in 2000 entitled *Data Management for Plant Collections, a Handbook of Best Practice.* The box on page 126 shows information taken from the section of the handbook dealing with the introduction of living material into the Garden. The handbook suggests standards for notes and information that should be recorded when bringing any plant material into the Collection, particularly when collecting wild plants in the field.

Making herbarium specimens

During a field trip, plant material is collected into polythene bags or placed temporarily into field presses during the day. Depending on the type of expedition and climatic conditions, it is either pressed periodically throughout the day or sorted, pressed and recorded once back in the camp. Working up the material ideally requires three people: one scribe for the Collecting Book, one 'presser' to number and prepare the specimens and one person to prepare the material in advance of the presser.

Plant material is arranged on a 'flimsy' (a thin sheet of paper or newspaper), so as to display as many important diagnostic features as possible, and at least one leaf is turned over to show the reverse side. Tall plants can be folded but excess leaves and other plant material are removed if the specimen is too thick and will not dry efficiently. Fleshy parts such as fruits and succulent

stems can be sliced for easier drying. Bulky structures, such as cones, are removed and dried separately. The flimsies are then stacked one upon another with a thick sheet of blotting paper between them and, once the stack is big enough (usually about 15 specimens) wooden presses are placed on the top and bottom and the stack tightened with belts. The specimens stay in the flimsies with their unique collection number until they are finally mounted back in the Herbarium.

Water from the specimens passes through the flimsies and is soaked up by the blotters. In temperate countries, the presses are usually placed overnight in a drying frame suspended over gentle heat. In warm, dry countries, drying frames are still required but can be supplemented by natural drying out in the sun or on a roof rack. Care needs to be taken to ensure that specimens are fully dry as fruits and flower heads often take longer to dry out than leaves.

If specimens are packed away damp they are likely to go mouldy and ruin other specimens packed with them.

In the humid tropics it is usually impossible to dry material properly and so a different technique is used. Specimens are collected into polythene bags, which are then partially filled with alcohol. This preserves the material long enough for it to be brought back to RBGE before it gets pressed and dries in the Garden's drying room. The only drawback with this technique is that the specimens turn black.

The details of all the plant material (herbarium and seed) are recorded in the expedition Collecting Book. Computers are increasingly being used in the field to record this information, which reduces the time needed back at Edinburgh as the notes can be downloaded directly into RBGE's plant records database.

Once fully dry, specimens are bundled in two dry blotters and tied firmly with string. The bundles are

put into clean plastic bags and sealed with tape to keep the specimens dry. The dry specimens are quite vulnerable to damage and therefore need to be stored carefully for the remainder of the fieldwork. Once back at RBGE, the specimens are frozen at −18°C for three days to kill any insects that might damage the material. Specimens are then mounted onto thick card, a label with all the collection information is added and the plants are then available for immediate research or stored for later usage.

left Placing plant specimens into a temporary field press in south-west China, 2002.

top right Specimens, firmly secured in presses, drying naturally in the sun. Atacama Desert, Chile, 2008.

middle right Heaters and drying frames are used to speed up the process in parts of the world where high humidity makes it difficult to dry and store field specimens. Nepal, 2001.

bottom right Staff unpacking bundles of dried herbarium specimens after an expedition to Nepal in 2001.

Equipment

The destination, duration and purpose of the fieldwork will dictate the type and amount of equipment taken. Camping adds enormously to the quantity of equipment required as does work in wet and/or cold countries. Where large quantities are required equipment is packed into trunks and airfreighted out in advance. If fieldwork requires visiting a country on a frequent basis the best option is to leave the equipment in the host country.

above Essential kit: a selection of fieldwork equipment including press, GPS recorder, camera equipment, first aid kit, compass, maps and insect repellent.

opposite Attention to detail: the list of equipment on the right has been adapted from the report to central and eastern Nepal 2001. The list shows the sheer amount of personal and plant-collecting equipment required for a typical Himalayan expedition of about five weeks mounted to collect seeds and herbarium specimens for research and reference purposes.

Group Expedition Kit – essential

Group first aid kit
GPS (Garmin GPS 12 –
at least 2, we took 3)
AA batteries (lots for GPS,
digital camera, pocket
computer, head torches, etc,
mainly Duracel M3 Ultra)
A4 hardback notebooks (2)
for Collection book, 30cm ruler
and protective plastic bag
Selection of marker pens, ballpoint
pens, pencils and rubbers
Sticky tape (2 rolls) and
packing tape (2 rolls)
Nylon string (2 balls)
Rubber bands for moss packets
Pre-folded moss packets and
A4 paper to fold more
Plastic bags: large sacks for
collecting and protecting
dried specimens (35)
A4 collecting bags (lots)
A5 and smaller bags for
DNA samples etc (lots)
Cotton bird bags for seeds
and fruits (100)
Small envelopes for dried seed
Small greaseproof sachets
for dried seed
Silica gel for DNA samples (3kg)
Prefolded thin paper
flimsies (2,000)
Blotting paper (250 sheets)
Aluminium corrugates (100)
Press ends (12 pairs)
Press straps (20 pairs)
Aluminium drying frame
(collapsible) and tools for assembly
Foil-lined groundsheets to
wrap drying frame (2 x
Lifesystems mountain sheet)
Large bulldog clips (20)
for drying frame
Kerosene wick stoves (2)
(bought in Kathmandu)
Spare wicks (bought
in Kathmandu)
Kerosene and jerry cans
(bought in Kathmandu)
Matches and gas lighters
Heat dissipators (2)
for kerosene stoves
Metal trunks (2) (bought in
Kathmandu and big enough
to take three press ends)
Large kit bags (4) (bought
in Kathmandu)
Padlocks for kit bags

Group Expedition Kit – desirable

Digital Camera (Nikon
Coolpix 995)
Memory cards for digital camera
(4 x 64Mb CompactFlash cards)
Electronic data recorder
(Psion pocket computer
with Palmtec case)
World receiver shortband radio
Clothes pegs

Group Expedition Kit – luxury

Whisky
Pepper grinder
Good quality instant coffee
Dundee cake
Dried fruit and nuts

Individual's Day Kit – essential

Rucksack (large 'day sack')
Rucksack cover
Rucksack liner (strong plastic
bag or large collecting bag)
Shoulder bag or bum bag (for
notebook, GPS, collecting bags etc.)
Hand lens
Altimeter (we took analogue
Thommen 6000m)
Compass (Silva)
Wristwatch
Field notebook
Digging equipment (ice axe
and/or stout trowel)
Penknife (preferably lockable)
Secateurs/Clippers
Camera equipment (35mm
SLR) and spare film
Water bottle (Platypus
or Sigg flask)
Small first aid kit

Individual's Day Kit – optional

Umbrella (non-collapsible)
Money pouch or wallet
Trekking poles or walking stick

Individual's Main Kit – essential

Kit bag (robust zippered
90l 'soft' bag)
Padlock (for kit bag)
Plastic sandwich boxes (2 or
3) (for small items such as
medical supplies, wash kit, etc)
Small dissecting kit (forceps,
fine scissors, blades, etc)
Multi function pocket
tool or good penknife
Sewing kit and safety pins
Notebook for diary
Photographic film (we took
20 x 36 exposure 100ASA
slide films per person)

Camera batteries
Head torch and spare
batteries and bulbs
Thermorest (standard,
if possible long)
Sleeping bag (3 season, down
preferred for cold, dry conditions)
Sleeping bag liner
Nylon drying line
Trek food (museli bars,
dextrose tablets, flapjacks)
Maps
Guidebooks (extracts
or photocopies)
Botanical reference material
(local Floras, checklists etc)
Reading material
Address lists

Individual's Main Kit – optional

Small calculator
Travel pillow
Insulated trek mug
Pack of playing cards

Individual's Clothing – essential

Trail or trekking walking
boots (well broken in)
Camp shoes (trainers or
walking sandals)
Waterproof shell overcoat
Waterproof overtrousers
Leech leggings (2 pairs)
Walking socks (3 pairs, including
short socks for hot days)
Sunhat
Sun glasses
Warm hat or balaclava
Scarf or neck wrap
Bandana (2)
Gloves (preferably windproof)
Down waistcoat or jacket
T-shirts (3) (wicking
fabric preferred)
Long-sleeved shirts (3)
to cover arms in sun
Base layer fleece or sweat shirt
Mid layer fleece
(windblock preferred)
Thermal long johns
or fleece trousers
Lightweight trekking trousers
(3 pairs) (belts as necessary)
Lightweight shorts (2 pairs)
Underpants (5 pairs)
Thin socks or liner socks (2 pairs)
Handkerchiefs (3)

Individual's Clothing – optional/luxury

Down bootees (great for
high altitude camps!)

Down trousers
Fingerless mitts (for fiddly work
in the cold) or spare gloves
Spare pair of walking socks
Swimming trunks

Individual's Clothing and travel kit to leave at base (Kathmandu)

Smart trousers
Smart shirt
Lightweight smart shoes
Underpants (2 pairs)
Socks (2 pairs)
Handkerchief (1)
Eye mask and neck support
(for sleeping on planes)
Universal sink plug

Individual's Toiletries – essential

Ear plugs (for travel and
noisy lodges)
Toilet roll
Wet wipes (2 x packs)
Small or medium towel (quick-
dry trek towel recommended)
Face cloth
Razor and spare blades (if
you are going to shave!)
Shaving brush/brushless
shaving cream/shaving oil
Soap and soap dish
Toothbrush and toothpaste (50ml)
Dental floss
Cotton buds
Shampoo (sachets or
screw-top bottle)
Deodorant
Nail clippers
Hairbrush or comb
Travel clothes wash
Water purifying tablets
Hand cream
Vaseline (small tub)
Lip balm
Sun cream
Insect repellent (including
Autan sticks™ – to protect
against leeches)
Vitamin C tablets
Knee supports
Blister kit
Plasters (various) and
micropore tape
Medicines

Individual's Travellers documents

Travellers cheques
Cash (including some US$)
Passport (with visa)
Copies of passport and visa
Spare passport-size photographs
Credit cards
Business cards

How the Garden grows

DEVELOPING AND MAINTAINING THE LIVING COLLECTION

From laboratory to mountainside, from microscope to tractor or chainsaw – growing and managing the huge diversity of plants in the Living Collection demands an equally varied range of human skills from the gardeners and scientists of RBGE. Strength and sensitivity are required in equal measure as Garden work ranges from soil analysis to tree climbing. [*And sometimes a pair of waders comes in handy too.*]

right A member of staff pruning the vanilla orchid (*Vanilla imperialis*).

Quarantine

Just as some animals have to undergo a period of quarantine when entering a country in case they are carrying contagious diseases, so the same is true of plants. RBGE has its own dedicated Quarantine House where all plants arriving from overseas must go for inspection.

It is standard practice to isolate newly collected material in the Quarantine House for a minimum period of at least three months after the initial inspection to be sure that it is free from pests and diseases.

Diseased plants are destroyed immediately and plants from the same expedition that appear to be disease-free are grown on within the isolation of the Quarantine House for a stipulated period of time. Only once they have been inspected again, not just by the Garden's pathologist but also by Government staff, can they be released for cultivation in the Collection. These regulations and procedures generally apply to plants or plant parts (such as roots, bulbs and cuttings) only and not to seeds, which can go straight to the openly accessible propagation houses.

Whenever possible, dormant organs such as tubers are brought into growth in isolation to establish that they are free from pests, diseases and viruses. This is especially important for material originating from outside the EU. Symptoms often only emerge several months into the isolation period, once the material is in active growth.

The hygiene regime inside the Quarantine House has to be rigorous to ensure that conditions remain as pest- and disease-free as possible. Tools, such as secateurs and pruning knives, are sterilised between cuts, and especially between different plants. Plant pots and accessories such as canes and labels are always new. Wooden canes are never used as they are difficult to sterilise completely and can aid the transmission of pests, such as red spider mites, root mealy bugs and fungal pathogens.

It is important to realise that humans can be the most serious vectors of pests and diseases. This applies especially to isolation areas – staff wear lab coats to reduce transfer via clothing – and to the movement of staff between isolation areas and other growing or displaying areas. Staff always wash their hands or wear gloves between handling new batches of material and other collections.

Those working in the Quarantine House need to demonstrate constant vigilance, as

well as attention to detail for careful cultivation. 'Good housekeeping' is critical – tidiness, weed control, cleanliness and the swift removal of any pests or diseased plants are all important and help the smooth progress of newly collected plant material through the House and into the Collection.

As a general rule, importing soil from third countries outside the EU is prohibited, except when soil is brought in for research purposes under a separate import licence. Any soil adhering to roots is generally washed off before transportation back to Edinburgh. As an alternative to soil, inert, unused or sterile materials such as coir, perlite, sphagnum moss or damp newspaper can be used to keep roots alive in transit. Where soil is essential for the survival of the plants in transit, small quantities attached to plants are permitted, but the more soil involved the greater the risk of importing associated pests and diseases.

Although there is free trade and therefore free movement of plant material between EU member states, a plant passport is required for the movement of some plants within and between member states. Additional conditions may be applied for particular species to enter what are termed 'protected zones', because of the risks that associated pests or diseases can pose to the food industry, forestry, horticulture and the environment in general. Records of plant passports and associated documentation must be retained for tracking purposes in the event of an outbreak. The list of plant material for which plant passports are required has recently increased to take into account new disease threats such as *Phytophthora ramorum* (Sudden Oak Death or Ramorum Blight).

The additional work required by these plant health and bureaucratic necessities adds considerably to the time, cost and responsibilities of collecting and managing an internationally focused plant collection. However, with the threat of accidentally importing potentially devastating pests and diseases that could spread into the countryside or infect commercial crops, it is essential that botanic gardens take plant health and quarantine issues very seriously.

Other sources of acquiring plants

Just as gardeners have always exchanged plants amongst themselves, botanic gardens throughout history have exchanged plants with each other. When one garden has a

above After inspection newly arrived plants are potted up or cuttings taken – often into an open compost promoting free drainage around roots – and then kept in a high-humidity environment to reduce water loss from leaves. Plants are inspected daily and any showing signs of disease or pest infestation are removed and destroyed immediately. After at least three months in quarantine (longer for cacti, succulents and bulbs), plants must be given the 'all clear' from the Garden's pathologist before release to the Living Collection. Government inspectors also make frequent visits to check for imported pests and diseases and ensure that the Garden is abiding by Government regulations. Here, tuberous roots of gingers in the family Zingiberaceae, collected in Sulawesi, have been potted up.

surplus of a particular plant, it makes sense to gift or exchange it – indeed some gardens circulate lists of available material, known as *indices semina*. The only difference in this procedure today is the current need for an agreement between the two gardens to ensure that they are conforming with the regulations of the Convention on Biological Diversity. RBGE occasionally obtains seed from *indices semina* but the main focus is always to try and obtain wild source material, which only some *indices semina* contain. RBGE's policy is that under some circumstances *indices semina* seed may be utilised, such as when a species cannot be obtained by other means, or when the primary purpose is for display, interpretation or education. However, the problem with *indices semina* is that since the material is often of garden, or possibly hybrid, origin, it may be unverified and without data and so is of little or no value to botanical research.

The Garden occasionally buys in some plant material from commercial nurseries, especially for predominantly ornamental areas such as the Herbaceous Border or the Queen Mother's Memorial Garden, or for a project that has to be planted at short notice. In general, however, RBGE tries to avoid purchasing plants from nurseries because of the Garden's overriding remit to be a scientific plant collection containing wild collected plants with full documentation. Though some specialist nurseries sell very interesting wild origin plants with a certain amount of collecting data, the Garden can never be absolutely sure that such material has been collected with the agreement of the country of origin and within the regulations of the CBD.

Seed sowing and propagation

Most plant material enters the Living Collection in the form of seeds (there were 1,283 accessions in 2010, accounting for 60 per cent of all types of incoming material), but bulbs, live plants and other propagules are important sources of material too. By whatever method material enters the Collection, it needs to be handled or processed in some way so that it can be recorded, propagated, grown on and then planted out in the Gardens or glasshouses.

Fieldwork to collect seed in northern hemisphere countries takes place in the autumn, and since there is no seasonal difference between the place of collecting and the Garden, the seeds can be sown fresh upon arrival. Germination is often quick as there are likely to be fewer dormancy

mechanisms in force than when the seed has been thoroughly dried in preparation for long-term storage (as opposed to the superficial drying possible on an expedition). The material is generally sown in containers and placed outside for 'natural' germination, which is often influenced and enhanced by the elements, especially cold or frosty periods. Little additional attention is required, and seedlings emerge slowly as temperatures rise in the spring.

Collecting seeds in the autumn season in the southern hemisphere means that material arrives during our late spring. Experience has shown that a flexible approach is best, and where possible a portion of material is sown fresh while the rest is carefully dried to reduce the moisture content, thereby enhancing storage, for later sowing. Spring-sown material is generally subjected to warmer, drier weather than autumn-sown material and therefore requires closer attention. Seed trays or pots are usually placed in moderately warmed,

closed cases, which are inspected daily as germination can be rapid.

Tropical material is sown fresh where possible and placed in warmed, closed cases and inspected every day. Tropical species, in particular, can be 'recalcitrant' – meaning that their seeds cannot tolerate the effects of drying. This can be a problem as seeds are usually dried down to a low moisture content in order to improve storage (this is one of the major drawbacks of seed banks). If seeds of recalcitrant species are dried, the embryos will die. Examples of well-known plants that are recalcitrant include avocado, mango and lychee. For this reason all tropical material is sown as soon as possible after its arrival at Edinburgh.

The straightforward technique of sowing seed fresh in the autumn soon after its arrival usually works well for 'normal' seed (technically called 'quiescent' seed), which only needs to imbibe water at an appropriate temperature to initiate germination. Unfortunately, this does not work for all

Dealing with stubborn germinators

Occasionally, all efforts to germinate seeds fail and the Garden may seek advice from the Forestry Commission's Alice Holt Research Station in Surrey. This research station has the expertise and equipment necessary for rigorous trials and analysis and so has been very successful in germinating a number of species that the Garden had found impossible to germinate. Such was the case with the threatened Chilean conifer *Prumnopitys andina* (top). This species propagates easily from cuttings but seed germination is very slow, erratic and difficult, so when the International Conifer Conservation Programme required seed-grown plants for one of its projects, the Garden asked staff at Alice Holt Lodge to help crack the problem.

After several trials a process was discovered which resulted in satisfactory germination. First the seeds had to be removed from the fleshy fruits; this was achieved by squashing, squeezing and rinsing. Then, to ensure there was no trace of any pulpy fruit material left (as it might contain a germination inhibitor) the seed was rinsed in running water at 15°C for 12 hours.

Preliminary trials using a nutcracker or hammer on the hard-coated seeds to release the inner embryo either bruised, squashed or completely flattened the contents of the seeds! Later, a more refined and delicate technique using a carefully controlled vice (centre left) was found to be successful and, once the seed was cracked, it was possible to prise away the seed coat and remove the embryo. Embryos were transferred to moist filter paper (bottom left) and incubated in conditions alternating daily, with 16 hours at 20°C in the dark followed by 8 hours at 30°C in the light.

Slowly, roots and shoots emerged and the resultant seedlings were potted up and grown on. They are now fine, healthy young trees some of which are growing in the grounds of Dunkeld Cathedral (bottom right).

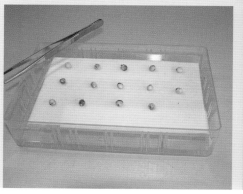

143

species. Seeds in all shapes, sizes and conditions are received directly from the field and some are difficult to germinate, especially if they have been dried and stored, because of various types of in-built dormancy mechanism. Dormancy is a condition where seeds will not germinate even when the environmental conditions (water, temperature and aeration) appear suitable for germination. Dormancy mechanisms in plants have evolved to prevent immediate germination and ensure that it takes place at the best time to ensure survival, for example, avoiding the middle of seasonally cold or dry periods.

Many species of plant have evolved different types of dormancy to promote their survival in their particular climate and habitat, and so gardeners have tried to mimic events in nature to break these mechanisms. At RBGE, numerous techniques to enhance

seed germination and break dormancy are followed, depending on the species. These include procedures such as chilling, or chilling followed by warming, exposure to light or total darkness, scratching the surface of a hard seed coat or immersion in different chemicals.

While seeds are the easiest and preferred method of collecting and transporting plant material, it will sometimes not be possible to collect them if it is the wrong time of year, or if there are none available. In these cases it may be appropriate to collect cuttings or seedlings. This is best done near the end of a visit so the young susceptible material can be rushed back to Edinburgh for immediate care in the Quarantine House. Likewise, for some species (in genera such as *Crocus*, *Begonia*, *Cyclamen* and *Iris*) it is possible to collect bulbs, tubers or rhizomes. This is often a successful alternative way

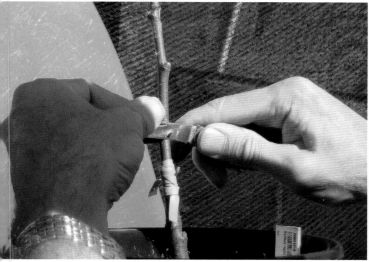

to introduce new plants to the Garden, although they are heavy to transport and so it is usually only possible and desirable to collect them in small numbers.

Grafting and budding are techniques used in the nursery stock industry to produce cultivars, mostly of ornamental trees such as rowan (*Sorbus*), crab apple (*Malus*) and cherry (*Prunus*). These techniques involve creating a union between a single bud or short piece of stem from one plant and the rootstock of another and can be useful where the cultivar does not produce a strong root system. It also allows for large numbers of plants to be produced from a small amount of plant material. Grafting and budding can have an application in botanic garden horticulture or conservation where woody plants have produced neither seeds nor suitable extension growth from which to make cuttings.

Fern cultivation

Ferns have been of special scientific and horticultural interest at RBGE for many years and their varied fronds contribute texture to each of the four Garden landscapes. Over 500 species are now maintained within the Collection and their cultivation has become one of our specialities.

Fern cultivation differs from that of seed-bearing plants because instead of seeds and flowers, ferns have a two-stage life cycle. When fern spores settle on a suitable damp surface, they germinate and develop into little heart-shaped plants known as gametophytes, which anchor to soil, bark or rock and produce both sperm and egg cells. The gametophytes are carefully tended in climate controlled growth cabinets; they need to be kept moist to allow the sperm from one gametophyte to travel to the egg of another but not overcrowded and so they

left Trees are normally planted out when they are about 1–1.5m tall, 2–3 years old and in a young and vigorous stage of growth as this promotes rapid establishment. A few trees are grown on to larger sizes in the Nursery for projects that require immediate visual impact – for ceremonial or commemorative plantings or where gardens are not yet ready to plant smaller trees. In these cases trees are grown in Air Pots™ (sometimes known as 'hedgehog pots') which promote a very fibrous root system.

are 'patched off' – little groups about half the size of a stamp are planted into new pots to allow the best conditions for fertilisation.

Once fertilised, the second stage called a sporophyte develops – this is the adult plant, which will grow fronds. When mature, spores will develop on the underside of the fronds and so the fern life cycle begins again. The time between sowing of spores and the appearance of sporophytes can be as little as three months but in one instance, for the filmy fern *Trichomanes reniforme*, it took fourteen years!

When the young sporophytes are large enough, they are pricked out in just the same way as flowering plants, and then potted up and grown on until they are ready to be planted out in the Garden.

Growing on and planting out

Once germinated, seedlings are pricked out, generally into individual pots, and kept in an environment appropriate to their natural habitat. Plants grown from cuttings, once rooted, are taken from the propagation house and potted up to join seed-grown plants in the glasshouse or whichever environment has a temperature regime best suited to their needs. Temperate plants, for instance, are grown on in an unheated

glasshouse where the doors are kept open during the day to encourage maximum air movement. Depending on the species in question and its speed of growth, temperate plants are progressively potted up and eventually leave the closely monitored care of the nursery glasshouse for a polythene tunnel or shade house. Some woody material is grown on in the open ground in nursery stock rows for development into specimen trees or shrubs, but this is the exception rather than the rule as the Nursery is not geared up to handling large plants grown in the open ground.

Luckily RBGE's four Gardens are not susceptible to either the vandalism or the stressful environmental conditions associated with many urban parks and so it is possible to plant out stock that is young and still in a vigorous phase of growth. Wherever possible, trees are planted out into the Garden at two to three years old and when they are less than one and a half metres tall. This not only eradicates the need to grow material on to a large size in the Nursery, which requires space and staff time, but also ensures a high rate of survival and vigorous early growth.

Hardy bulbs, rhizomes and corms are potted up, grown on and planted out, often in mesh pots, to ensure that, even if they

above left At one time the Garden used an enormous range of potting composts to provide a bespoke mixture for almost every species. Fewer composts are used now and most are bought in. However some are still specially made, particularly for tropical plants. Compost mixes use ingredients such as different grades of bark and loam from a loam stack for aquatics. Additives include perlite, pumice, sand, hen grit, moss, leaf mould, charcoal, limestone and fertilisers.

above right Staff planting a Wollemi pine (*Wollemia nobilis*). This tree, received as a 200mm plant from Australia, was grown on until large enough to plant in the Garden. Following *Collection Policy* guidelines, care is taken to maintain a multi-age and variable-density tree canopy while enhancing the landscape and diversity of the Collection.

1 *Trillium grandiflorum* 'Roseum' from woodlands of eastern North America grows in parts of Edinburgh's shaded Woodland Garden.

2 *Verbascum dumulosum* from Turkey demands good drainage and thrives in gravel on a raised bed.

3 Australian bottle brush tree (*Callistemon citrinus*) thrives at Logan if planted against a wall.

4 Large-leaved *R. macabeanum* needs substantial wind shelter.

5 A gravel mulch looks aesthetically right for the Rock Garden but organic mulches are best in woodland gardens.

6 The limestone wall in front of the Alpine House is ideal for calcareous plants.

7 Understorey woodland plants require dappled shade, provided by an overhead canopy.

8 Alpines suffer in Scotland's wet, mild winters. A pane of glass keeps them dry but not warm.

9 Lobster claw (*Clianthus puniceus*) from New Zealand survives outdoors at Logan.

10 Plunge planting in sand is for highly water sensitive alpines, like some species of *Helichrysum, Raoulia* and *Haastia*.

11 *Echium nervosum* from the Canary Islands grows outdoors only at Logan – the other Gardens are too wet or too cold.

12 Troughs allow bespoke composts and drainage to be created for specific groups of species.

---

above Adapting the microclimate. Edinburgh has a dry, cold, windy climate and the soil is sandy and acidic. Nevertheless, a surprising number of species can be cultivated because there are few extremes. Even with this favourable climate horticultural staff are able to 'bend' the microclimate to cultivate even more plants. The Alpine House, for instance, is designed to keep alpines dry over winter.

divide and spread laterally to the extent that they encroach on another species, those remaining within the pots are correctly labelled. This is especially important when grown close to other species of the same genus. Species of smaller and more delicate bulbs, such as smaller daffodils (*Narcissus*), *Fritillaria*, *Crocus* and *Muscari*, are often kept in pots in the Alpine House.

Tropical material is handled in a similar way and, depending on size and species, is potted up progressively until it is large enough to plant in one of the display glasshouses. Research material is generally kept in pots and the plants are maintained in this way until they reach an unmanageable size, after which they are propagated from cuttings where possible and regrown. This ensures that the same genetic entity is recycled through propagation, growing on, maturity and repropagation.

At one time the Garden used an enormous range of composts in which to cultivate plants – the theory being that almost every species required a dedicated 'special' compost in which to grow. A rationalisation down to a few standard mixes with a few variations as required occurred a number of years ago and this has been largely successful.

Close collaboration between Curators and the Nursery Manager ensures that plant material is tagged or earmarked for particular destinations early in its life and so can be grown in the quantities and quality required. Particular pot types, composts, formative pruning and cultivation requirements are discussed on a frequent basis until it is time for the plants to be collated ready for transport to one of RBGE's four Gardens.

Site selection and adaptation

Britain's moist, seasonal maritime climate allows the cultivation of an amazing number of species. Even so, gardeners are always trying to find ways to 'push' their garden's microclimate a little further so that they can grow just one more, special exotic. With RBGE's remit to grow a vast diversity of plants from different habitats all over the world, staff employ every available technique to adapt the soil or microclimate in order to accommodate the requirements of all these species.

RBGE is fortunate in having four Gardens

located in different parts of Scotland: cold, dry Edinburgh with thin sandy soils; wet, mild Benmore with humus-rich but often shallow soils; warm and wet Logan; and almost continental Dawyck with warm summers and very cold winters. These four climates are used to the full with the most appropriate species going to each site, as determined by the *Collection Policy*.

Yet within these four climates, a considerable amount of time is spent adapting the macro environments to create numerous micro environments. The subtle exploitation of the environmental conditions created by aspect, topography and canopy cover are all used to grow the maximum diversity of plants on each site. In addition to these natural features, Garden staff have created all sorts of man-made features such as walls, artificial windbreaks, glasshouses and irrigation as well as mulches and other soil ameliorations.

Aspect, shade and shelter are important considerations. Some plants, such as those from Mediterranean climates, thrive in south-facing warm sites and tolerate dry conditions and sandy soils but grow less satisfactorily in shade. Understorey forest herbs, on the other hand, grow best in the shade of trees or on the north side of buildings. Large-leaved rhododendrons, while revelling in the high rainfall of Benmore, need wind shelter, which is why all the classic west coast rhododendron gardens have substantial wind breaks. An appreciation of aspect also allows for an understanding of frost drainage and frost pockets.

Soils are just as important as temperature, aspect and shade, and there are all sorts of things that can be added to make soils more fertile, moisture retentive or free draining. Mulches of compost and leaf mould retain moisture, suppress weeds and look appropriate in woodland gardens while mulchings of gravel do the same job but look best in rock gardens. Amelioration (digging in or mixing) of the soil in the root zone (as opposed to mulching) with either grit or compost can improve draining or retain moisture. The soil's acidity or alkalinity can also be modified by the addition of limestone or peat.

left Despite their long life and dominant stature, trees require as much care and attention as other plants. They are also subjected to a risk and hazard evaluation process on a regular basis with branches or even whole trees being removed if necessary. Crown lifting and thinning help manipulate crown density and height. Here Benmore's skilled arboricultural staff are section felling a beech tree (*Fagus sylvatica*).

The skilled combination of these techniques enables RBGE's horticultural staff to grow an astonishingly diverse collection, with plants from many habitats and climates. For example, Edinburgh has a thin sandy soil and low rainfall, conditions that are completely unsuitable for growing rhododendrons, yet with automatic irrigation, the use of mulches and wind shelter they flourish here. Likewise, Edinburgh has an acidic soil with a pH of 5 yet calcicoles, which tolerate lime, such as some species in the genera *Gentiana*, *Draba*, *Dryas* and *Dianthus*, thrive in the limestone wall in front of the Alpine House. Inside the Alpine House, the cold, dry environment protects susceptible high-altitude alpines, adapted to surviving winter under snow, from our damp winters.

At Logan, *Clianthus puniceus*, a threatened leguminous species from New Zealand, survives outdoors on a sheltered sunny wall although it is generally considered to be tender, at least in northern Britain. Likewise species of *Echium* from the sunny Canaries, *Puya* from arid Chile and *Grevillea rosmarinifolia* from Australia, all grow luxuriantly in the suntraps created by the raised stone beds backed by tall stone walls. Benmore and Logan are both vulnerable to ferocious Atlantic storms, and so dense woodland and shelterbelts are essential to protect the Collection from wind damage. At Dawyck, with its exceptionally cold winters, only the hardiest plants can survive but an understanding of cold air drainage in the undulating topography and the careful selection of plants from Nepal, North America and northern Europe allows a remarkable diversity of species to flourish.

Cultivation and care

RBGE grows huge numbers of plants in its capacity as both a visitor attraction and a scientific resource. Maintaining almost 130,000 plants, gathered from 161 countries in 4 gardens and 22 glasshouses, is a considerable feat requiring trained staff, technical facilities and numerous environmental adaptations. Some plants are grown away from the public's gaze for scientific purposes only while others are cultivated purely for public enjoyment. The majority, though, have a dual function as plants both of research potential and for amenity. Whatever their intended purpose, all plants are treated with the greatest care and are skilfully cultivated, recorded, tracked, pruned, mulched, labelled and observed.

The cultivation requirements depend on the type and size of the plant in question. For a typical outdoor tree or shrub, the cultivation cycle in the first couple of years after planting out might include staking, formative pruning, irrigation, mulching and weed control, along with visual inspections for any pests or diseases. The post-establishment years, usually after two growing seasons, might include mulching, some irrigation in prolonged dry periods, pruning and weeding.

Once established, trees still need care and attention to maintain their health and vigour and on occasions will need arboricultural interventions such as crown thinning – the removal of a proportion of the branches to

Gardening skills and tasks:

1 Edging

2 Pruning

3 Potting

4 Lopping branches

5 Blowing leaves

6 Grass cutting

7 Top dressing gravel
   on paths using a
   power barrow

8 Application of residual
   herbicide

9 Pruning tropical plants

10 Scaffolding tower required
    for pruning tall trees

11 Scarifier

12 Blowing

13 Weeding

14 Planking

15 Back actor removing
    a small tree

reduce weight – or canopy lifting, where lower branches are removed to let more light reach plants below the canopy. Limb removal may be necessary where serious decay has taken place or where branches are misshapen, asymmetric or crowded.

For large, established trees in the Edinburgh Garden, a risk and hazard evaluation inspection for public safety reasons is required for every tree, every year. At the Regional Gardens, with more wooded landscapes, lower visitor numbers and fewer public paths than in Edinburgh, a sector by sector approach is adopted – a whole area is inspected as a block and, if necessary for safety reasons, can be cordoned off from the public.

For herbaceous plants, the attention needed after planting involves mulching, staking if necessary, weeding, deadheading during the season and cutting back at

the end. Pest and disease control may also be necessary, as may irrigation in prolonged dry periods and fertiliser applications. The plants are then split and divided once the crowns become congested, usually after five to seven years.

The cultivation requirements of indoor plants depend on whether they are planted out permanently in the display glasshouses or are kept in pots 'behind the scenes' in the research glasshouses. Those in the display glasshouses are treated similarly to outdoor trees and shrubs: after planting they are mulched, given fertilisers, pruned as necessary, weeded, kept within their allotted space and then repropagated and replanted if they become too big.

Pests and diseases can easily get out of control in heated glasshouses, especially in hot, humid tropical houses, and so more time is devoted to pest and disease control for glasshouse plants than outdoor plants. The

Garden adopts an 'integrated' approach to pest and disease control, which combines growing healthy vigorous plants in the first place, for example by using regular foliar feeds, 'good housekeeping' such as the removal of dead leaves or diseased stems, and the use of biological control (such as the predatory mite *Phytoseiulus persimilis* for red spider mite, the parasitic wasp *Aphidius colemani* for aphids or the fungal pathogen *Verticillium lecanii* for whitefly and mealy bug). If these measures fail, then there is careful use of insecticides and fungicides.

Plants in the research glasshouses range widely in their size and needs, from tiny *Streptocarpus* species that are grown from seed to maturity and then quickly resown or propagated by leaf cuttings, through to tall, leafy members of the ginger family (Zingiberaceae), which tower almost to the top of the glasshouse roof. There

are also woody tropical conifers being grown as part of the International Conifer Conservation Programme.

Every plant in the Living Collection has a story to tell – of its journey from the field to the Garden, whether as a seed or small plant, its careful cultivation in the Nursery and the constant attention to its needs as it grows and flourishes out in the Garden or in the glasshouses. The extraordinary diversity of the plants in RBGE's four Gardens are a result of the tenacity of the horticulturists and botanists who collected the material in the field, the skill of the propagation and Nursery staff who nurtured the seedlings and the dedication of the staff who care for and maintain the Collection year after year, in good weather and bad. Their hard work, dedication and immense skill has created one of the greatest Living Collections in the world.

RBGE adopts an integrated pest and disease management programme, using a combination of cultural, physical, biological and chemical techniques.

left Pesticides are used as a last resort when other techniques have failed. Here tropical plants are being sprayed with Abamectin cyclohexanol for an outbreak of red spider mite.

top right Biological control is used as often as possible and this photograph shows species of *Aphidoletes* being applied to *Capsicum chinense* 'Bhut Jolokia' (incidentally, one of the hottest chilli peppers there is) to control aphids.

bottom right Indoor plants are fed regularly during the growing season with liquid fertiliser.

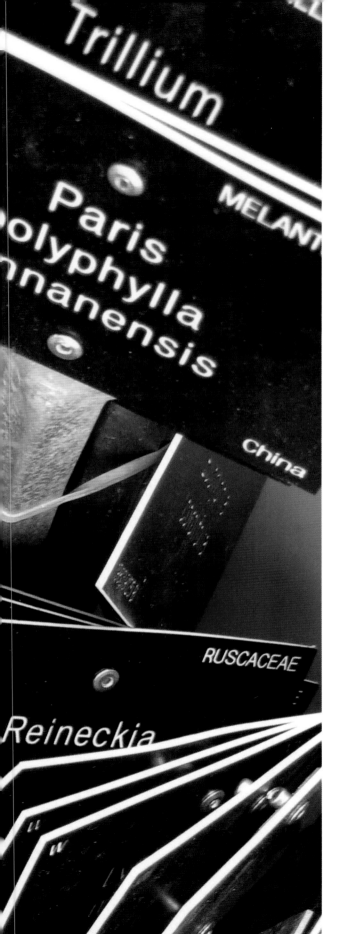

Keeping track

RECORDING THE LIVING COLLECTION

An efficient plant records system is crucial to the management and use of the Living Collection and is now regarded as an integral part of the cultivation process. The Garden needs to be able to locate, track and store information about every plant in the Collection in order to manage the plants efficiently. Keeping track is a complex and time-consuming task which is mostly invisible to visitors to the Garden. This diverse work involves accessioning, labelling and verifying the plants as well as stocktaking, auditing and publishing the plant records.

left Labels display the most important facts about a specimen such as its name, family and country of origin, while the accession number (on the bottom left of each label) links that specific collection of the species to the information held about it on the Garden's database. Replacing broken labels, ensuring they are in the right place – as well as ensuring the name is correct and up to date with the latest scientific treatment of the plants – is a time-consuming business and keeps one member of staff busy full time.

Records for living material at the Garden go back as far as 1810 in the form of hand-written books in which all plants were listed under the donor and the entire batch was given a single number. Comprehensive records in accession books survive for the latter half of the 19th and first half of the 20th centuries, handwritten from 1838 to 1957 and typed thereafter.

RBGE was one of the first botanic gardens in the world to computerise its Living Collection records. In 1969 the arduous task began of transferring information from accession books into a mainframe computer run by the Scottish Office Computer Services. For many years, the manual records were updated daily and the accumulated changes were sent on to the Scottish Office for incorporation into the database. Eventually the Garden acquired its first computer and this process was brought in house in 1983.

RBGE's current database system, BG-BASE, was adopted in 1990. Originally designed for the Arnold Arboretum of Harvard University, it has been widely used in the US. Prior to purchasing BG-BASE, the Garden was considering developing its own,

unique software package – a process quite common in botanic gardens at the time but fraught with all sorts of problems. However, BG-BASE offered a comprehensive, proven system which could handle, and integrate, all of the Garden's Collections records, including living and preserved collections. It meant that large amounts of information, in many formats, could be included and that individual plants from within one accession could be tracked and managed. The comprehensive nature of BG-BASE means it can process the Garden's historic records as well as current information.

The system has since been developed and many more modules have been installed, allowing RBGE staff to fully integrate the information management of its living and preserved plant collections. Since 1994, the plant records activities have been devolved so that members of the horticultural staff can update the programme directly.

More recent developments in computing have transformed some of the record keeping process. Improved capacity and the increasing availability of wireless networking means that staff now have real-time access to the computer databases while working

in the Garden or in the field. All four Gardens have been digitally surveyed since 2004 and a mapping package now makes it possible to print plans showing the location of plants on the ground at each Garden. If, for instance, our Arboriculture Supervisor wants a plan showing the location of all our oaks (*Quercus*) or our archivist wants a plan showing the location of all the plants connected to a plant collector such as George Forrest, this can now be done at the press of a button.

Accessioning

When a plant or seed enters the Living Collection, from whatever source, it is assigned a unique accession number that remains associated with it forever. The person logging the material into the records system, usually the Nursery Manager or the Plant Records Officer, requires the following minimum information which would have been recorded when the plant or seed was collected or acquired: type of material (such as seed, bulb or plant), source of the plant and collector name, number and associated details. This process is called accessioning. Once a plant has been given this number,

despite perhaps looking very similar to other plants or seeds, it can be located, sorted, stored or processed in any way, even if it does not yet have a name.

Accession numbers at RBGE are eight digits long. The first four represent the year the material was accessioned and the last four the sequential item number in that year. Labels in the Garden often show both the accession number and the collector's number, which is assigned by a collector to a specimen while in the field.

In some cases, however, the first four digits of the accession number clearly do not reflect the year in which the plant material was brought into the Collection. For example, when RBGE introduced computerised plant records in 1969, all existing plants were then accessioned for the first time and those whose year of introduction was unknown were simply assigned 1969 as the first four digits of their accession number.

The accession number can refer to a single plant, a batch of seeds, a group of cuttings or other associated material such as leaf samples or herbarium specimens. For wild collected plant material, an accession represents a plant or plants that are all of

right RBGE was one of the earliest botanic gardens to computerise plant records with the task of transferring the data onto a computer beginning in 1969. In 1990 the Garden adopted *BG-BASE*, a relational database application utilising a variable-length and multiple value field data structure (giving great flexibility not usually found in simpler fixed-length field systems). *BG-BASE* is a large system, able to track and integrate all of the Garden's Collections, encompassing some 7,000 fields of data spread over 200 database tables.

the same taxon and collected from one site by the same collector on the same day. Subsequent collections from the same site on different dates, even if by the same collector, are regarded as new accessions.

Accession numbers are unique and the same number is not reused even if a plant is lost or has died. Only in specific circumstances are accession numbers changed. If they are, this is called 're-accessioning' and this can happen when, for instance, a plant has lost its accession number and that number cannot be accurately identified from existing records. Batches of plant material originally accessioned under a single number and understood at a later date to be a mixture of different taxa would also be re-accessioned.

In general, cuttings, grafts and layers taken from existing accessions are not re-accessioned but are given the same accession number as their parent because they are of the same genetic stock, having been propagated asexually. However, while the accession number would remain the same, a note would be made in the record that this was second-generation material. Almost all seed derived from accessioned material should be re-accessioned before it is sown because, having been propagated sexually, it will be of a different genetic make-up from either of its parents, information that anyone wanting to study the plants needs to know.

Seed or spores are also re-accessioned if taken out of storage for use in the Living Collection some time after a first part of the original batch was sown. For example, when spores of the oblong woodsia (*Woodsia ilvensis*) were collected from the wild in 1996, the accession number 19961234 was allocated at the time. A portion was sown

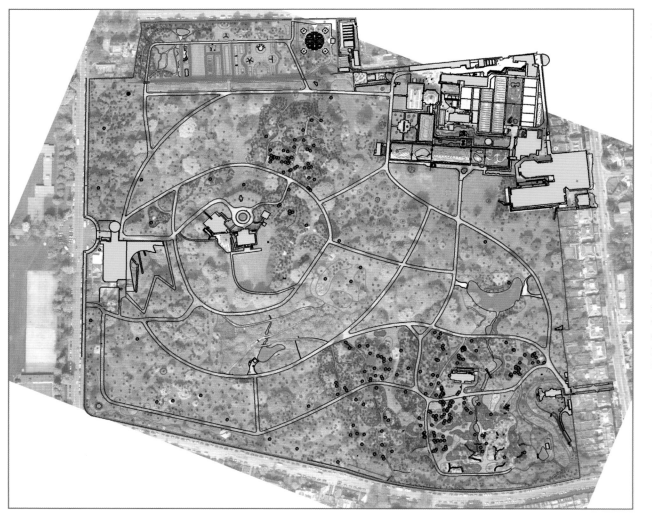

left Mapping. In recent years RBGE has been further refining and extending its plant records capability. From 2003 to 2008 each of the four Gardens was digitally surveyed. Buildings, boundaries, beds, borders, paths, ponds and contours – along with all the dominant trees – were surveyed and recorded. In subsequent years Garden staff added the remaining trees and most of the shrubs. Not only does RBGE now have detailed plans to use in planning and project work but it means that the position of trees and shrubs, from any combination of information held in the database, can be plotted and printed. The black dots on the plan show the position of plants collected by George Forrest.

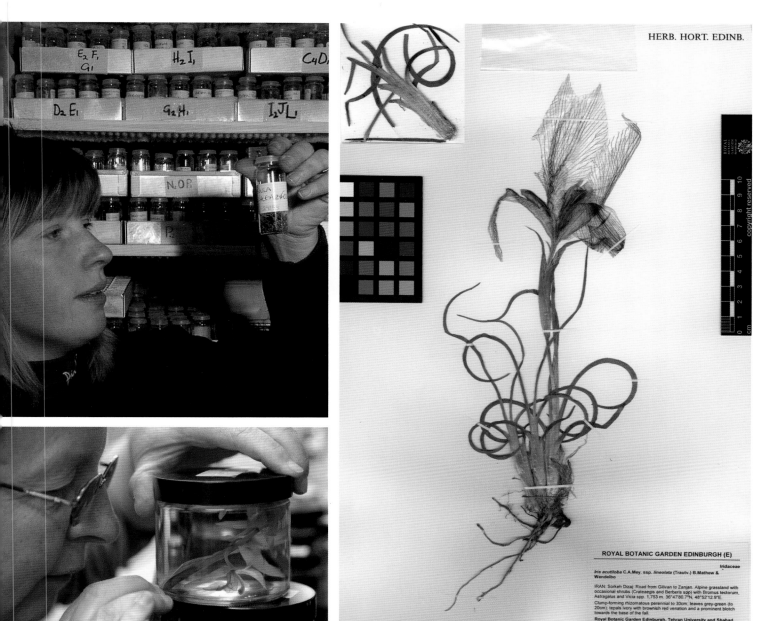

ROYAL BOTANIC GARDEN EDINBURGH (E)

Iridaceae

Iris acutiloba C.A.Mey. ssp. *lineolata* (Trautv.) B.Mathew &
Wendelbo

IRAN: Sorkeh Dizaj. Road from Gilivan to Zanjan. Alpine grassland with
occasional shrubs (Crataeegis and Berberis spp) with Bromus tectorum,
Astragalus and Vicia spp. 1,753 m. 36°47'80.7"N, 48°52'12.9"E.
Clump-forming rhizomatous perennial to 30cm: leaves grey-green (to
20cm); tepals ivory with brownish red venation and a prominent blotch
towards the base of the fall.
Royal Botanic Garden Edinburgh, Tehran University and Shahad
University (ETU) Joint Expedition to Iran May 2005 No. 1
Collected with Tony Miller, John Mitchell, David Rae, Mehdi Zarrei

11 May 2005

ROYAL BOTANIC GARDEN
EDINBURGH

and the remainder stored in the germplasm
collection (seed bank). In 1999, as part of a
research project investigating the longevity of
the spores in storage, it was decided to sow
part of the remaining batch. Therefore the
second set of spores was re-accessioned as
19991236 in order to make the recording and
monitoring of the new batch of plants easier
and to avoid confusion with the older plants.
However, there will be a note in the records
system tagging this new accession number
back to the original accession number
(and the information attached to it) so that
anyone working on the newer batch of plants
can understand what has taken place.

Qualifiers are the letters that immediately follow the accession number and are
used to record and track the individual
plants or germplasm that make up a single
accession. For instance, an accession might
contain fifty or more seeds of which forty
might germinate and from which ten are
grown on for planting in the Garden. Each
of these ten is assigned a qualifier (A–J)
when planted out in the Garden. Qualifiers
are essential for tracking and recording
individual plants, particularly woody plants
from within the accession for those scientists who require that level of refinement, as
well as horticulture managers who need to

above The great benefit
of *BG-BASE* is its ability to
integrate the records of all
the Garden's Collections with
the advantage of a common
nomenclatural standard and
easy cross-referencing. This
means that records of collections such as the germplasm
collection (seed bank, top
left), spirit collection (bottom
left), herbarium collections
of pressed and preserved
plants stored in the Herbarium
(such as the specimen of *Iris
acutiloba*, right) and Living
Collection are all stored on
one integrated system.

track the propagation history and subsequent performance of individual plants as opposed to a batch of seedlings.

Lineage numbers allow staff to track a plant's past progression from generation to generation when required. A lineage number is the accession number under which the genetic material first came into the Garden; if the accession number changes for any reason in subsequent generations (for example, if the plant is propagated asexually) then the lineage number allows reference back to the original accession number, with its attached information.

While some of these protocols can seem very detailed and possibly a little confusing, they are at the very heart of any botanic garden's work as they allow the most detailed scrutiny and recording of an individual plant's collection, propagation, repropagation and planting out. They enable those who want to use plants for research or conservation to have all the details they need at their disposal. It is only if a plant has been collected in the wild, and these sorts

of meticulous records are kept, that plants cultivated in botanic gardens can be of real value for scientific research.

Plant labels

Labels inform the public about a plant's name and origin and are critically important for both the scientific integrity and the general understanding of the Collection.

'Primary labels' are intended for the public to view in the Garden. These are the engraved labels with white text on a black background, in various shapes and sizes. These small labels can tell a lot about the plant's story, with a very comprehensive range of information including family, genus, species, hybrid with parents, subspecies, variety or cultivar, common name, collector's name and number, country of origin and, sometimes, conservation status, such as T for threatened. The use of letters on labels is also used to indicate if a plant has been verified (V) or is of wild origin (W).

The use of common names on labels causes all sorts of confusion because of the

Plant labels

1 Plant family.

2 Genus.

3 Species.

4 Subspecies, variety or, if in inverted commas, cultivar name. The word 'cultivar' is the correct term for what used to be known as garden varieties.

5 Vernacular or common name.

6 Collector's name and number: this identifies the individual who gathered the specimen from the wild along with their numbering system.

7 Country or geographic area to which the plant is native.

8 Plant accession number in the RBGE plant records system.

9 Qualifier (indicated by the letter) to plant accession number identifying individual plants or groups of plants within an accession.

10 Special codes. Letters or symbols indicate the status of the plant:

T The plant is threatened or endangered in the wild (show as * on older labels).

V Verified: the identity of the pla has been confirmed by an ex

W Plant of wild origin, acquired directly from the wild (showr + on older labels).

① EUCRYPHIACEAE

Eucryphia ②
× nymansensis ③
'Nymansay' ④
(cordifolia × glutinosa)

⑧
1957.8371 0

W• ⑩   OLEACEA

⑤ "Chinese Ash"
Fraxinus
chinensis

⑥
Lancaster 427
1981.0152 A ⑨

⑦ Chir

below Engraved labels with white text on black background are known as primary labels and display all the basic information about the plant in question.

bottom RBGE is increasingly experimenting with secondary labels which are embossed onto metal. These can carry more information but are less visually clear than primary labels. Ideally, trees at least would have both types of label with the primary label used mostly by the public, students and general users and secondary labels used by staff and also for double security in case the primary label is lost or broken.

variety of names given to plants, including regional and national differences. For instance, *Campanula rotundifolia* is commonly called harebell in England but bluebell in Scotland, whereas the plant known as bluebell in England is *Hyacinthoides non-scripta*. Similarly, *Galium aparine* is called either goosegrass or cleavers depending on which part of the country you are in. There are international differences too; for instance *Liriodendron tulipifera* is known as the tulip tree in Britain but as yellow poplar in its native eastern USA. It is for these reasons that scientists use the universally agreed Latin nomenclature.

Despite these problems the use of common names is important for the public and the lack of them is frequently cited as an issue in visitor surveys. Although not universally adopted throughout the Living Collection, common names are generally used for British native plants, all European and North American trees and shrubs (using British common names) and trees and shrubs from other countries where they

are considered to be 'of garden or horticultural interest' (using the *European Garden Flora* as a guide and again using British common names).

The Garden also requires other types of labels. During the propagation and growing on stages of plant development, and before a plant reaches its final destination in the Garden, nursery labels are used which have a bar code and essential information. Secondary labels are usually thin metal strips displaying only accession number and qualifier and are intended as a back-up in case the primary label is lost. These are particularly useful at Benmore and Dawyck on woody plants.

Verification

All who use the Living Collection, whether research staff, students or visitors, rely on plant names being correct and so the Garden spends significant resources on the process of 'verification', confirming that the name given to a plant in the Living Collection is correct and, where it is found

below Nursery labels are used during the propagation stages of a plant's life before it is planted out in the Garden. As well as name and accession number they include information which is important to Nursery staff, such as sowing date. They also incorporate a bar code, used in stocktaking and inventory work. (The red dot signifies that the germination date has been recorded.)

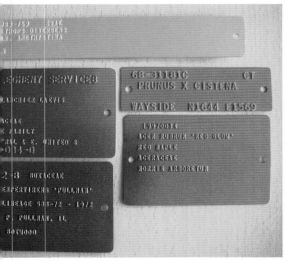

not to be, accurately identifying and naming the plant. The process also involves checking that the name selected is correct according to a recent monograph (a book describing all the species in a family) and that it is valid according to the *International Code of Botanical Nomenclature* or the *International Code of Nomenclature for Cultivated Plants*, the documents laying down the rules for naming plants.

The need for verification arises on many different occasions. For instance, a newly collected plant may be flowering for the first time, and so reliable identification may not have been possible before. Or there may be a plant which was only determined to genus level at the time of collection. It may also be just a keen-eyed observation that an existing plant in the Collection does not seem to conform to the characteristics expected of a particular species.

Accessions in need of verification are selected by horticultural staff (sometimes in conjunction with a specific request by a member of the scientific staff) and taken to scientific staff according to their areas of geographic or taxonomic speciality.

There are two steps involved in the verification process. The first is to confirm the existing name under which the plant originally came into the Garden or, if unknown, to determine its identity. The second step is to ensure that the name selected is valid and accepted by the wider botanical community. While this might sound straightforward, it can be complicated and time-consuming and so the Garden prioritises groups of plants such as those that are the focus of research, for example the gingers (Zingiberaceae), *Rhododendron* and plants from accessions that have not yet been identified.

On completing the verification process, the Plant Records Officer is given the information for updating the records, and a new label can be initiated if necessary. At this time the Verification Level is also entered. This is a number that indicates the level of authority that the person concerned has in verifying the plant.

−1 = known to be incorrect
0 = unchecked
1 = determined by comparison with living plants or by a horticulturist
2 = determined by a taxonomist using herbarium facilities
3 = speciality of taxonomists
4 = type or authentic material

Despite the time and commitment devoted to verification, as explained above, 77 per cent of the Living Collection remained unverified in 2000. This did not mean that the plants were incorrectly labelled, simply that their identity had not been thoroughly checked by an expert. With the uncertainty

left Staff involved in verifying the names of plants. Verification is the process of checking that the label is displaying the correct name – using the most up to date taxonomic treatment of the plant in question – and, if not, determining the correct name. It is a slow and time-consuming process but it is vital to ensure that the Collection is accurately named.

right If there is uncertainty about the name applied to a specimen, that plant, along with a verification form, such as the one shown, is taken to the most appropriate person who then examines the plant in detail and decides its correct name. Plant specimens are also frequently sent to experts around the world.

Verifying *Alnus* and *Acer* using the targeted approach – a personal account by Martin Gardner

During the summer of 2000, we took samples of all 35 accessions of the alder genus (*Alnus*) from the Living Collection, which we pressed and put to one side. Then, one day during the winter we spread them all out on a long table in the Herbarium along with representative examples of correctly identified material plus all the necessary reference books. Once we had familiarised ourselves with the key diagnostic features, the actual process of verifying all 35 accessions was relatively quick and easy. As well as a lot of new information, it resulted in the discovery that four species were incorrectly named. Further refinement of the protocols for this approach to verification was undertaken in 2001 when we took on the much larger genus *Acer* (the maples). We verified 242 accessions of *Acer* (shown here) using the same process and the exercise resulted in the confirmation that the vast majority (197) of the accessions were correctly named. However, it resulted in name changes for 35

accessions through identification and re-identification and a further 10 accessions had their names changed because of synonymy. Two accessions were found to be dead and four could not be verified because of a lack of diagnostic characters.

This approach to verification proved to be much more

time-efficient than the rather *ad hoc* approach normally used. The collection and drying of the 302 specimens that we gathered took around 10 days in total. The review of taxonomic literature took another day, and the curation of the existing reference material in the Herbarium a further ten days. The actual

verification of the 302 specimens once the background work had been completed took 3 weeks. When trying to verify a single, unfamiliar species 'cold', where similar specimens and reference books have to be sourced and key diagnostic features understood, it can easily take two or more hours to track down the correct name.

that students could be studying plants that were incorrectly labelled or that scientists could be sent the wrong plant, a decision was taken to make a concerted effort to drive up the number of verified plants and for that purpose a new, 'genus-based' approach was devised.

The 'genus-based' approach targets a single genus that is in need of comprehensive verification. Over an allotted period of time (usually one to three years) all accessions of that genus are collected from the Living Collection and pressed, the data is noted and a herbarium specimen made. This approach has proved to be far more time-efficient than the traditional, rather *ad hoc* system because, after the plants are pressed during the growing season,

they can all be spread out together, along with all the necessary reference books and existing herbarium specimens. Then, having familiarised oneself with the important diagnostic characters, it can be relatively straightforward to name large numbers of plants in a comparatively short time.

Data Capture Project

The flow of living material through botanic gardens has been described as 'the river of biodiversity'. Though the public might see little change in the content of a botanic garden from one year to the next, the professional staff know only too well that there is a constant cycle of germination, growth and death taking place. The places occupied by dead plants are quickly taken by new plants,

The walnut family – a case study for the Data Capture Project – a personal account by Natacha Frachon

The purpose of the Data Capture Project is to maximise the use, in the long term, of the Living Collection. In the knowledge that plants in the Collection inevitably die, and that many of them do so before ever being used for any scientific or conservation purpose, a decision was made to try to 'capture' as much information as possible about them while they are still alive. Following detailed discussions of the protocols and plant record requirements, a plan emerged to make a herbarium specimen, photograph and take a leaf sample stored in silica gel for later DNA extraction of 1,000 wild origin accessions each year. To test how this would work in practice, we used the walnut family (Juglandaceae) for the trial which was undertaken in 2007.

At that time the Living Collection had 26 accessions of Juglandaceae – 12 of garden (G) origin and 14 of wild (W) origin and these were represented by 30 plants growing in RBGE's Gardens, such as *Pterocarya fraxinifolia* seen left, with flowers below. A further seven accessions of wild provenance were in propagation. We collected herbarium specimens from 28 plants (1 plant at Benmore was dying and another, at Dawyck, was too young to take material from). We verified all 28 as part of the process and, as a result, 15 names were deemed to be 'good' and remained unchanged, 4 names were changed (2 because of technical reasons in applying the correct name and 2 had been incorrectly named) and 8 names remained uncertain as further diagnostic features were required, such as the shape of nuts and buds. At the same time as we made the herbarium specimen, we took a photograph and a leaf sample to put in silica gel to keep in long-term storage for future DNA extraction and analysis.

Although the three elements of the Data Capture Project were then complete (herbarium specimen, photograph and leaf sample) we continued with further curatorial work, ensuring that the section in the Herbarium holding Juglandaceae was comprehensively curated and arranged according to the latest literature. We also undertook a curatorial survey to review the management and content of the Living Collection. We found, for instance, that there were two accessions of *Pterocarya rhoifolia*, one of garden origin, in excellent condition and adding huge value to the landscape, and the other of wild origin, in poor condition and struggling to grow under the canopy of a sycamore, with the top of its leader branch broken. We recommended that this situation receive immediate attention. For the future, we also noted that we have two accessions of *Juglans ailantifolia* var. *cordiformis* which is a species from Japan, both of garden origin and both in poor condition. We suggested that the next expedition to Japan bring back wild collected seed.

either of the same species or completely different, and so the integrity of the landscape is maintained while individual plants come and go.

While some of these plants might not be so significant, others will have been collected in the wild or are plants of potential scientific interest. Wild origin plants are expensive to collect and their death marks a potential wasted opportunity for research. The purpose of the Data Capture Project is to capture or harvest as much information as possible about every wild origin plant

in the Living Collection while the plants are alive, to ensure that essential information or data has been captured and stored from them should they die.

The aim of the Data Capture Project is to store at least one herbarium specimen, a leaf sample for future DNA extraction and research and a photograph of each wild origin plant in the Living Collection. It is a time-consuming task, but the effort will greatly enhance the value of the Living Collection. The current target is to capture 1,000 specimens in this way each year so that over the years this accumulated data will build up into a considerable body of knowledge ready for use by the scientific and conservation communities both within the Garden and outside.

Stocktaking

Living plants bring their own challenges when collecting information about them. Most museum or scientific collections are fairly static, but the living specimens in a botanic garden tend to be moved, die, self-propagate or survive underground as bulbs or herbaceous plant roots for periods. Therefore stocktaking is crucial: this is the process of physically checking that the plants listed in the database as existing in the Living Collection really do exist on the ground (also known as 'ground truthing'). It is an important process because it is frustrating for all concerned if, having checked a database or list to see if a plant is present in the Collection, a subsequent search for the material in the Garden leads to a label with a dead or absent plant beside it.

The process usually involves producing a list of all the plants in an area or bed from the database and then checking to see if the plants are still alive. At the same time as simply recording whether they are alive, it is also worth noting their condition and any other factors such as particularly good flowering, fruiting or autumn colour. Ideally, all plants would be checked each year but the time needed to find and check all the plants, and return to the bed at different times of the year to check for bulbs or herbaceous plants, invariably precludes this. It is really only practical to take stock of the entire Collection around once in five years and especially before publication of a catalogue of the Living Collection; ideally, however, herbaceous plants and short-lived perennials would be checked once every two to three years.

In recent years staff have been investigating technologies to speed up the process of stocktaking using computing equipment which would enable them to have live, real-time access to the database while working in the Garden. With the increasing availability of wireless networking and improved capacity in portable computers,

right Ground truthing: stocktaking, sometimes also known as 'ground truthing', is the process of checking that the plants listed in the database actually exist – and are living – on the ground. It's another slow and laborious process but absolutely essential as any users of the Living Collection have to know that we really do cultivate the plants we claim to have.

right and bottom Modern technology is helping to speed up the process of stocktaking and RBGE will continue to invest in modern applications to improve the speed and efficiency of the process. The equipment shown here is a ruggedised tablet PC which can run the database software out in the Garden making recourse to paper lists and data entry back in the office a thing of the past – records can be updated in the Garden in real time.

The results of the Collections audit for the genus *Alnus*

Five-yearly audit of alder (*Alnus*). The figures show the number of wild origin taxa, total number of plants, number of wild origin accessions and total number of accessions for the years shown in the left hand column. The right hand column shows the percentage of wild origin for each of the years. The bottom two rows show percentage changes over two different time periods. Audits of key families and genera allow Curators to track changes in the Collection over time.

Year	Total no. of plants	Taxa	Total accessions	Wild origin accessions	% of wild origin
1990	91	20	48	31	65%
1995	131	25	63	46	73%
2001	316	23	59	42	71%
2007	317	22	58	45	78%
2011	422	24	74	62	84%
% increase 1990–2011	+ 364%	+ 20%	+ 54%	+ 100%	
% increase 2007–2011	+ 33%	+ 9%	+ 28%	+ 38%	

it should be possible for staff to search the database, read outputs and update records while they are out in the Garden.

The Collections audit

The Garden recently undertook an audit to track changes in taxon numbers (a taxon refers to any plant name, be it species, subspecies, cultivar or any other rank below species level and is therefore useful as a collective term) and accession numbers for important families and genera over recent years. This exercise proved to be of great value and so the Garden is now developing various annual and five-yearly reviews of the Collection. A number of key indicators, which collectively give an impression of the 'health' of the Collection, are being developed and, where necessary, targets are being set for areas considered to be weak or in need of special attention. This process can be likened to an annual health check where a doctor might take the blood pressure, cholesterol level and weight of a patient to give an impression of their overall health. If suitable checks are selected, such a process can give an indication of the Living Collection's 'fitness for purpose' – in other words the suitability of the Collection to meet the Garden's needs.

To improve the Collection's 'health', the Garden has now set specific targets to guide the development of the Living Collection. For example, targets have been set for a 2 per cent increase in verification and for 2,000 new accessions each year. One important target is to steadily increase the wild origin percentage of the collection by 1 per cent per year. Wild origin material is important because it means that a plant's exact locality and background are known and therefore it can be traced back to its exact locality in the wild, making it more useful and important for scientific study. As many trees and shrubs have long life spans, change in this area is necessarily slow and so it will take years of steady progress to increase the overall percentage of wild origin material. However in the period 2003–2010, the percentage of wild origin material has risen from 53 per cent to 56 per cent, a small but significant increase.

Since 1990, the total number of taxa, individual plants, wild accessions and all accessions have been gathered every five years for selected genera and families in which the Garden has a special interest. From these figures, trends in terms of percentage increase or decrease over a period of time can be generated, such as an increase or decrease in wild origin accession compared to 'all origin accession'. This process proved to be valuable and will continue to be undertaken every five years.

The audit gave Curators factual analytical data on the performance of key families and genera for the first time. It showed that the Collections were increasing in many key areas and that some families and genera had grown very considerably while others had remained static and a few had contracted. The audit demonstrated that in most cases where there had been a dramatic increase, the family or genus had been subject to concentrated collecting for a specific purpose or research project. Conversely, where numbers had remained static or had declined, it was because there was now less research emphasis on the family or genus.

In addition to these targets and audit, the Garden collates and publishes the total number of families, genera, species, taxa and accessions each year, both for the historic record and also for review and analysis. While quality and value are ultimately more important than the simple size of a Collection, the overall number of plants does give an easily calculated measure of change. The ongoing size of a collection is a result of the number of new accessions compared to the number of deaths, and so to get a clear understanding of how a Collection is developing it is worth tracking annual accessions, annual deaths and total numbers.

At RBGE annual accessions were in the range of 3,000 to 4,000 during the 1980s and early 1990s. However, reductions in funding for expeditions and in the number of Nursery staff along with the additional complications of accessing new plant material because of the Convention on Biological Diversity (CBD) had the effect of reducing new accessions to fewer than 1,000 in 2002. As a result of a review of this situation, it was decided that 2,000 new accessions per year would be a reasonable and achievable target. With this in mind, the appropriate staff have been encouraged and trained and new sources of external funding have been sought. The results of this concerted effort have reversed the declining trend with 1,955, 2,570 and 1,784 new accessions being brought into the Collection for the three years from 2007.

below A page from the 1778 *Catalogue* showing plants grown at the time.

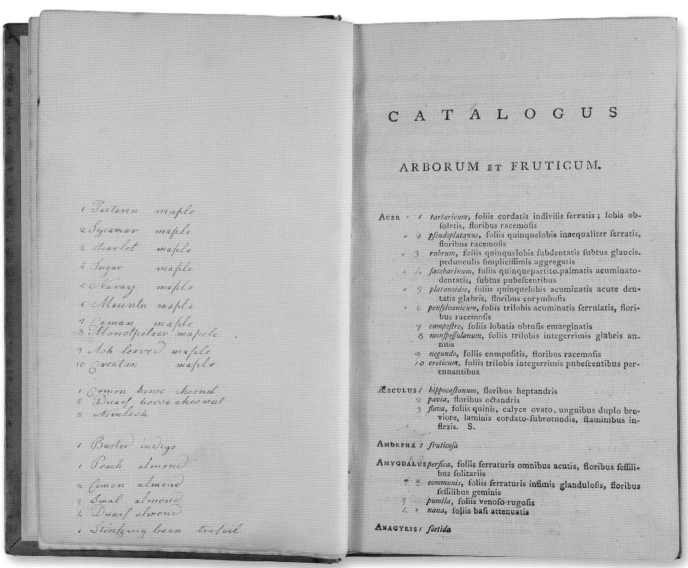

THE LIVING COLLECTION | 169

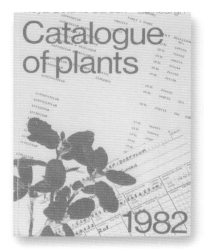
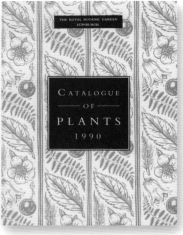
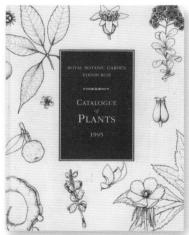
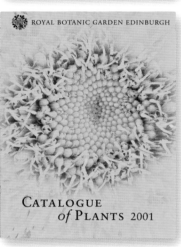

Reasonably high numbers of plant deaths in botanic gardens are to be expected for many reasons. The size and complexity of botanic garden living collections inevitably mean that they will include annuals, short-lived perennials, plants on the edge of hardiness and plants that are very demanding to grow and maintain. Unless figures are abnormally high, staff should not worry about high death rates as they are not an index of failure! However, where there are concerns about the quality of plant care it can be interesting to compare death rates for individual collections against others or one area of the Garden against another.

Publishing the Living Collection

Most botanic gardens have catalogues of their living collections, consisting of lists of plants growing in the garden concerned for a particular year. They are compiled for reference and archive purposes. While early catalogues usually contained only lists of plant species, more recent catalogues provide introductory information such as garden history and collection statistics and then further information adjacent to each taxon such as accession number, collector code and number and location within the Garden.

The very first RBGE catalogue dates from 1683, when James Sutherland, Intendant of the Physic Garden in Edinburgh, published *Hortus Medicus Edinburgensis* with the subtitle *or, A Catalogue of Plants in the Physic Garden at Edinburgh; containing Their most proper Latin and English names; With an English Alphabetical Index*. This was the first publication of its kind in Scotland and it signalled the Scottish entry into the established fraternity of European botanical gardens. Further catalogues were published in 1712, 1716, 1740, 1775 and 1778 followed by some departmental lists in 1896. After that, a few *ad hoc* lists were produced but no comprehensive list, as we understand it today, of the entire Collection.

Eight catalogues have been published

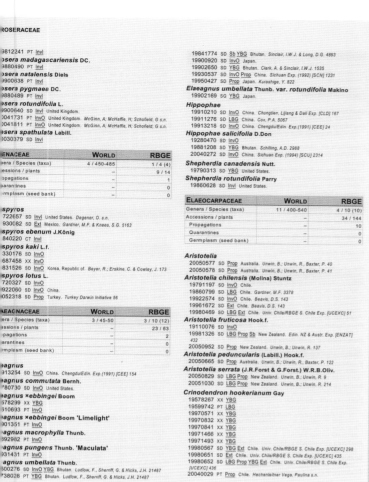

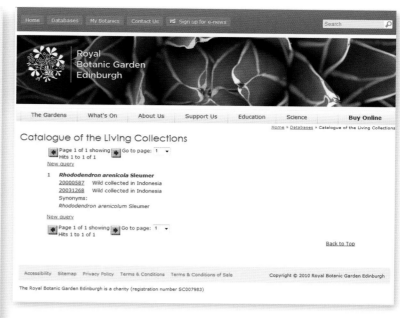

left A page from the 2006 catalogue showing the layout which includes a listing of every accession of each taxon held in the Collection, along with the Garden in which it is cultivated, its mode of entry (e.g. seed, plant, cutting or bulb) and the collector or expedition on which it was collected.

above Entry screen of the web-based catalogue. For many users a web-based catalogue is more convenient and up to date than a hard copy, published every five years. The web-based version is updated daily and available to all: public, students and research workers around the world.

at regular intervals since 1974, with a new standard set in the recent editions of 2001 and 2006. As well as overviews of Garden information and statistics, this is a listing of all the plants held in the Living Collection of the Royal Botanic Garden Edinburgh, including accession number, collector number (where appropriate), the Garden in which they grow and country of origin.

Plant records, while fascinating from a historical or archival point of view, are made to be used and so the Garden makes every effort to disseminate information as widely as possible. As well as the hard copy catalogues, the plant records information is all available on the RBGE website and this online version is updated daily.

The whole plant recording process has come a long way from the first handwritten ledgers and record cards to the present use of hand-held, real-time computing devices in the Garden. The techniques used to record and disseminate information have evolved rapidly with recent advances in computing technology.

Already we are electronically mapping the collection as an as an aid to management and public enquiry, and arboricultural staff are experimenting with the use of the latest mobile phone technology to record their assessments while they are at work high in the trees. Our processes and procedures, including verification, stocktaking, labelling and auditing, will all continue to develop, taking advantage of new progress in technology to ensure our records are as detailed, accurate and accessible as possible. This work – tracking every stage of a plant's life and development – is at the heart of managing a Living Collection. This is what botanic garden horticulture is about, and what separates us from other types of garden.

3
Using and enjoying the Collection

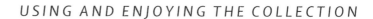

Safeguarding the future

RESEARCH AND CONSERVATION

"Measuring progress towards halting the decline in biodiversity is a key international obligation which cannot be achieved without baseline knowledge of biodiversity. Creating baselines and monitoring change is dependent upon the availability of taxonomic expertise across the range of living organisms. Systematic biology underpins our understanding of the natural world. A decline in taxonomy and systematics in the UK would directly and indirectly impact on the Government's ability to deliver across a wide range of policy goals. In view of the Committee's concern that demand for taxonomic skills will exceed supply, stimulating the recruitment of new researchers and new volunteers is vitally important."
House of Lords report, 2008

left Cataloguing the world's plants: RBGE's research is centred on plant systematics which is the science of classifying, describing and naming organisms both living and extinct, and of investigating the relationships between each other and with their environment. The purpose is to catalogue and describe the world's plants and provide a sound scientific basis for plant conservation. Here a scientist at the Garden works on plants in the ginger family (Zingiberaceae) using a combination of resources including herbarium specimens, plants from the Living Collection and plant parts stored in alcohol.

top Fieldwork: the ultimate purpose of the Garden's research is to conserve plants in their natural habitat. To do this, scientists need to undertake fieldwork to gather specimens and data for research and record population changes over time. Once a species' biology is fully understood, conservation strategies can be developed and implemented.

The different elements of our mission statement 'Exploring and explaining the world of plants for a better future' are intertwined and in sequence. We first have to explore the world of plants through our research to be able to explain it through our work in education and public engagement. Our endeavours in both exploring and explaining contribute to conservation and so to a better future for the world's plantlife and for humankind.

The main driver of our research at RBGE is the need to underpin conservation with well-founded evidence, because of the serious threat posed to the world's plantlife by global environmental change and mass extinction. We explore the world of plants by classifying and seeking to understand their distribution, evolution, diversity and conservation. Our scientists provide baseline botanical data by investigating the evolutionary processes that have given rise to botanical diversity and through practical involvement in plant conservation.

Documenting and describing the living world

Working in the UK and around the world, RBGE scientists look at the fundamental questions of what species there are and how they are distributed. The heart of our work is in systematic biology, the science of classifying, describing and naming organisms and of investigating how organisms relate to each other, how they have evolved through time and how they function ecologically, biochemically and physiologically.

The terms 'systematics' and 'taxonomy' are often used interchangeably, but strictly speaking systematics is the scientific study of the diversity of all organisms and of any and all relations between them, and taxonomy is that part of systematics that deals with the process of naming and classifying plants.

In descriptive taxonomy, the botanist describes and illustrates the species using

below left Staff working in the molecular laboratories using polymerase chain reaction (PCR) techniques to support evolution and conservation research. PCR is technique used to amplify region of DNA, generating thousands to millions of copies of a particular DNA sequence.

below right The Garden's traditional scientific speciality is taxonomy – the classification an naming of organisms. Ideally, classifications reflect both the organism's evolutionary history and its morphological appearanc (its shape and structure). Here, a member of staff is studying a member of the family Sapotacea (a family containing about 800 species of tropical evergreen tree

opposite Plant classification ha traditionally been based mainly on floral parts (like the number c petals and sepals) and the arrang ment of these parts relative to ea other. These four images illustra small selection of the diverse flo structures found in nature.

top left *Tacca chantrieri* in the plant family Taccaceae.
top right *Lilium primulinum* var. *ochraceum* in the plant famil Liliaceae.
bottom left *Paris japonica* in th plant family Trilliaceae.
bottom right *Gunnera manica* the plant family Gunneraceae.

above The art of science: as well as preserved and living plants, text books and illustrations also have a role in plant description and classification. Line drawings and full colour illustrations can help in the process of identification and skilled artists are able to incorporate the fronts, sides and reverse sides of flowers and leaves in their illustrations along with flowers in bud, partly and fully opened, plus fruits and seeds – attributes that are seldom found on one herbarium specimen or a living plant.

left *Fascicularia bicolor* ssp. *canaliculata*, from Chile.

right *Mitraria coccinea*, also from Chile.

reference literature and plant specimens – preserved or living – to study their structure, looking at the size, shape and arrangement of leaves and especially the reproductive parts (usually flowers). It is important to preserve as a herbarium specimen the actual material on which this descriptive account is based. This is known as a 'type specimen' and is the first example of a species to be officially described and named – the physical evidence for that name. Type specimens are usually dried and pressed and carefully looked after in the Herbarium as the international reference for a particular species and can be used to resolve any subsequent dispute over how the name was originally applied.

The procedure for naming is laid down in a set of internationally agreed rules, the *International Code of Botanical Nomenclature* for wild plants and the *International Code of Nomenclature for Cultivated Plants*, also known as the

Horticultural Code. Ideally, classifications reflect both the organism's evolutionary history and its appearance, but in many instances there are complications that confound this. Organisms that look the same, or share similar features, may have resulted from morphological convergence of different evolutionary lines and in fact be unrelated.

Through this systematic research we are working towards a global inventory of plant species. This is vital work – amassing the core knowledge that is essential to conserve biodiversity. Whether working with species and habitats in an ecosystem, or setting up a forest reserve for mountain gorillas, it is essential to know what plants are there. In the face of unprecedented habitat loss, we may be the last generation fortunate enough to have the opportunity to explore fully the diversity of life on our planet. Yet conservation work is impeded by a lack of knowledge; for example, more than 80 per cent of the large tree species in the Amazon

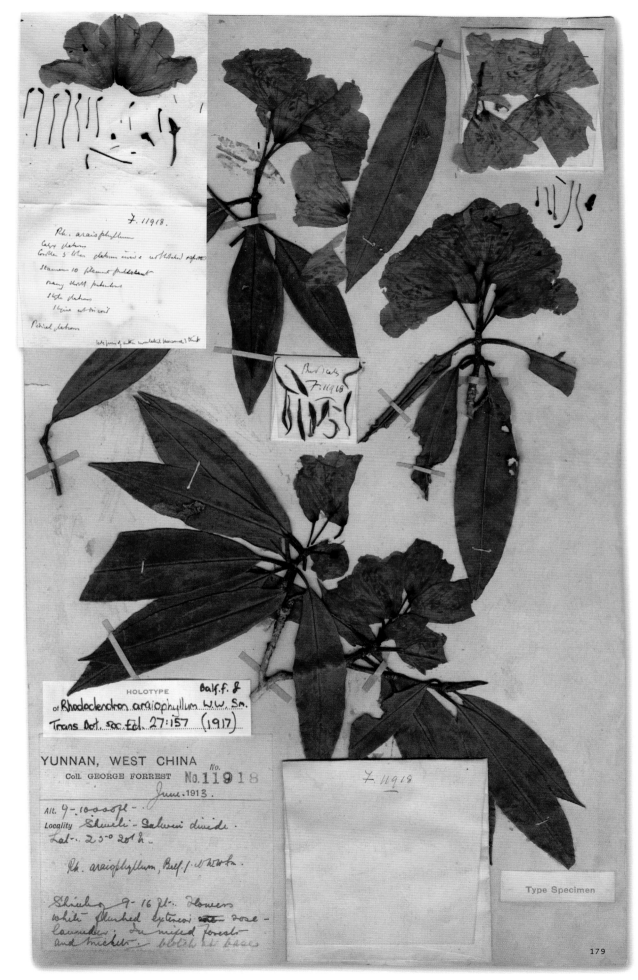

right Preserving history: the type specimen of *Rhododendron araiophyllum*, collected by George Forrest in June 1913 at 9,000–10,000 feet in Yunnan, south-west China. This is the actual specimen on which the formal description was based and is therefore described as a type specimen. The RBGE Herbarium holds about 300,000 type specimens which are curated particularly carefully as their importance lies in the ability of future research workers being able to study the actual specimen upon which the description and name were based.

Looking closer

Botanical research has been transformed in recent years by technological developments. The traditional image of a botanist is probably of someone out in the field with a hand lens, returning to study the plant under a low-powered microscope. This approach still has a role to play, but both the potential and the techniques of botanical research have been greatly enhanced by advances in technology. For example, the development of the transmission electron microscope (TEM), which can magnify up to 1,000,000 times, allows the detailed study of individual cells and cell contents. At RBGE it is used to study cell organisation and to examine the effects of disease and fungal infections. The scanning electron microscope (SEM), which can magnify up to 900,000 times, now provides superb images of the surface of individual cells and has been used extensively at RBGE, for example in studying the scales on rhododendron leaves as part of their description and classification. As a result of these developments a new range and detail of plant structures has been revealed, down to the level of pollen and spores, which adds a wealth of new data to the investigation of plant diversity.

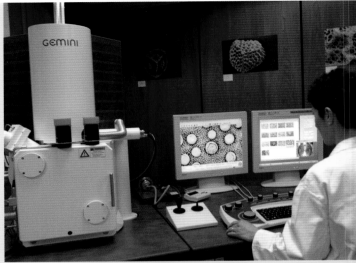

top The Garden's scanning electron microscope.

bottom left Underside of *Rhododendron anthopogon* var. *hypenanthum* leaf.

bottom right Underside of *Rhododendron grande* leaf.

---

below The task of cataloguing and describing the world's plants is not finished. Many species await discovery but may be lost before they can be named. New generations of plant systematists are required to help complete the task but are in short supply. RBGE's MSc in the Biodiversity and Taxonomy of Plants is helping to address this worldwide shortage. Here, students are engaged in a field trip to Belize.

rain forest have not been described. For non-flowering plants and fungi, the descriptive task remains immense – new species are still regularly being discovered even in well-documented regions such as Europe and North America and large areas of the world remain completely undocumented in botanical terms. The world's major botanic gardens have the range of resources and staff expertise that is needed to undertake large-scale systematic studies of depth and quality. We are working to speed up this urgent research by training new taxonomists and harnessing new technologies as well as through local, national and international collaboration.

Floras

Much of our research in systematics is focused on the creation of Floras – comprehensive descriptions of all plants within a defined geographic area such as a region, country or continent. RBGE has a long-standing history of, and an internationally renowned reputation for, floristic research, especially in the Himalayan region and Asia. The Garden leads major floristic projects around the world, often in areas that are rich in plantlife but where natural habitats are threatened. The purpose of a Flora is to allow the user to identify accurately any plant within the area concerned and to obtain a synthesis of current knowledge of that plant. Floras, therefore, are a vital resource for naturalists, researchers, conservationists, foresters and policy makers.

Writing a Flora is a large-scale, long-term commitment, often undertaken in partnership with institutions from many countries. The *Flora of Bhutan*, for example, includes

9 volumes and was written over a 26-year period, while the *Flora of Turkey* took 35 years, from the initial volume in 1965 to the final, 11th volume in 2000. This long time span is due to the extent of fieldwork and the depth of research required to create a comprehensive account. The botanists working on Floras need to be skilled in diverse techniques and disciplines. As well as descriptive taxonomy and the process of naming plants, the work requires expertise in biodiversity informatics – the science of the analysis, handling and storage of biological information. Staff use resources such as existing literature, live plant collections, herbarium specimens and specialist equipment such as the scanning electron microscope (SEM).

Understanding evolution

Plant evolution is the process by which genetic changes have taken place in populations of plants over millions of years in response to environmental changes. Evolution has resulted in the formation of new species and, usually, an increase in complexity. RBGE is exploring how plants are related and how they evolved, addressing issues such as why some plants are restricted in their distribution while others are widely distributed and the evolutionary

The *Flora of Bhutan*

In 1975 the Royal Government of Bhutan commissioned RBGE to produce the first-ever *Flora of Bhutan*. At the time, the Garden had a number of historic herbarium specimens from Bhutan but they were not at all comprehensive and the geographical coverage was patchy. Therefore before any accounts could be written, extensive fieldwork was required by RBGE botanists David Long and Andrew Grierson, the authors, along with Henry Noltie, of most of the *Flora*. This was at a time when Bhutan was just opening up to outsiders after decades of isolation and the two botanists had a privileged insight into this beautiful Buddhist country during the early years of the project.

The first of the nine volumes of the *Flora* was published in 1982 and the last, covering orchids, in 2002. The *Flora* is considered a vital reference work for botanists, foresters, academics and those engaged in conservation in Bhutan as it is the only publication covering all of the plants in the country. Its value was particularly appreciated after the signing of the Convention on Biological Diversity (CBD). All signatory countries were then charged with developing National Biodiversity Action Plans, and the *Flora* was essential for Bhutan to fulfil this commitment.

top left Republic of Congo: staff from the Garden are working to document the flora and with local colleagues training the next generation of botanists.

top right Staff at Edinburgh working on the *Flora of Arabia* which covers all the countries in the Arabian peninsula. It is a collaborative projects involving the countries concerned and has been in progress since the 1980s.

relationships between species. This research is fundamental to our understanding and appreciation of biodiversity. For example, it is used for predicting how plant species will respond to future climate change and to identify relatives of useful species which could be used as new genetic resources. The Garden's scientists make full use of new technological and intellectual developments to evolutionary and genetic research. Their challenge is to take new approaches that may previously have been used only in the context of a few organisms and to apply them to a wider sample of global diversity, from tiny diatoms and lichens in Scotland to the huge trees that dominate the world's forests.

In recent years, the focus of laboratory-based research at RBGE has been on the utilisation of DNA sequences to investigate the evolutionary relationships between

plants at all levels – from orders to families to species and populations within species. This has been accompanied by the development of powerful new tools to analyse and compare the huge sets of data that can be captured by DNA sequencing. The study of evolutionary relationships (phylogenetics) and the classification of the resulting information (cladistics) have contributed to major reorganisations of the classification of plants.

DNA 'barcoding' is a particularly exciting concept within the field of molecular ecology, which is concerned with applying genetic techniques to the study of biodiversity. RBGE has led a large-scale international collaboration to identify a suitable 'barcode' – a defined section of DNA that can be used to identify plants. This pioneering technique was already used to identify animal species, but applying the process to plants was far more difficult because of the complexity

opposite DNA fingerprinting gel showing genetic differences among individuals (each 'column' represents an individual). DNA fingerprinting is used to measure the amount of genetic diversity within populations.

below left Staff in the molecular laboratories preparing samples for DNA sequencing and then loading them into a DNA sequencer. DNA sequencing technologies have been used at RBGE since the late 1990s as a complementary research approach to traditional morphological plant description.

below right Barcoding life: a research assistant prepares DNA samples of a liverwort in connection with the DNA Barcoding of Life project for DNA sequencing.

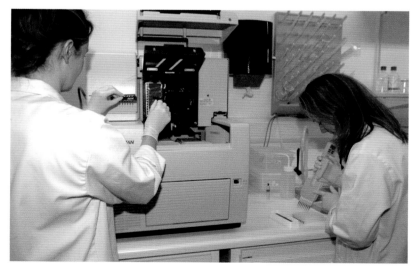

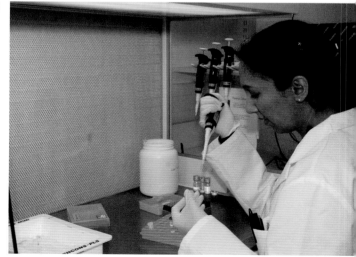

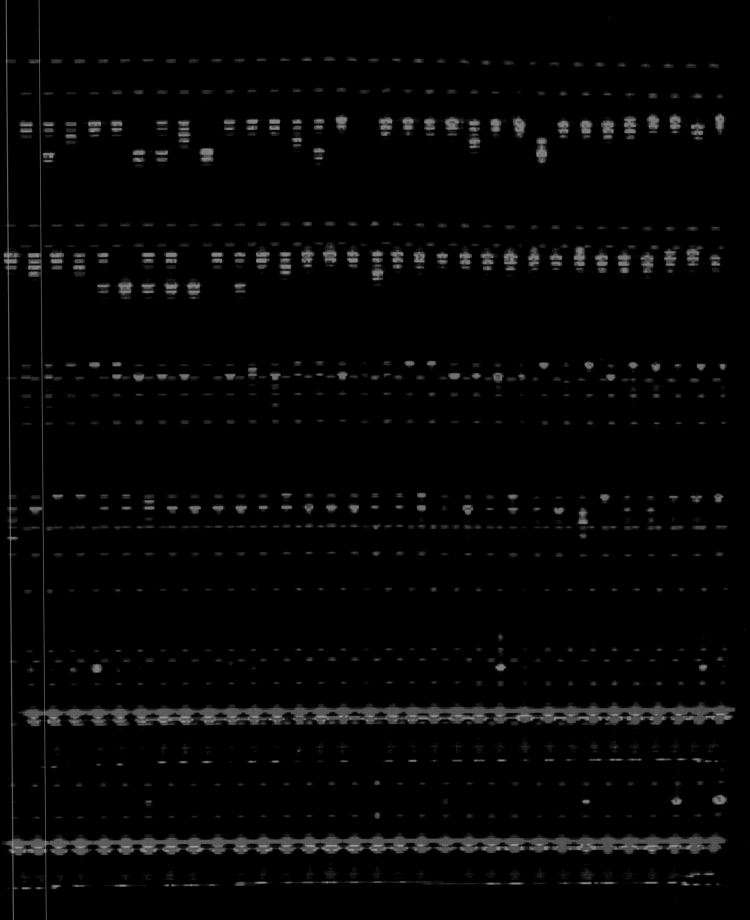

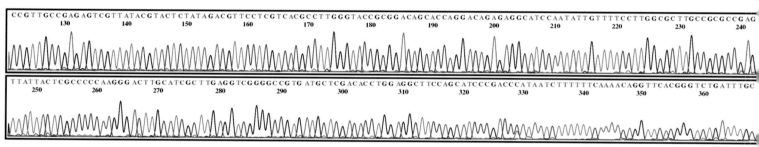

above Liverworts and other bryophytes, such as mosses and hornworts, can be slow to identify by traditional morphological techniques so make ideal 'guinea pigs' for modern molecular taxonomic approaches. The RBGE has a long-established reputation in bryophyte research and this continues through participation in the international Barcoding of Life project.

left *Saccogyna viticulosa*

right The viking prongwort from Scotland and Norway, a recently recognised species of *Herbertus*, still awaiting a formal Latin name, discovered during the Barcoding of Life trials.

below Sequence output from the viking prongwort.

of plant genetics and the complications of hybridisation, where plants can pass DNA between species. After a challenging process, involving 52 scientists in 10 countries over 4 years, a technique was agreed which can accurately identify approximately 70 per cent of plant species. This is a starting point from which the project can begin to build a vast, global reference library of plant DNA, against which DNA sequences from unknown plants can be queried and identified.

This technique does not replace traditional skills but enhances the speed and accuracy of the identification process, adding a powerful new tool to the taxonomist's tool kit. As it only requires minute samples of plant material, DNA barcoding could be invaluable for identifying illegal trade in endangered species or for forensic investigations. Above all, it could become a vital tool

in conservation, especially in areas where a shortage of specialist identification skills hampers efforts to protect plantlife.

UK liverworts (small, flat, green plants that are related to mosses) have been selected as 'guinea pigs' for initial DNA barcoding trials as the Garden has extremely comprehensive herbarium collections as well as long-established expertise in these plants. Liverworts are ideal candidates as they are difficult to identify by traditional methods, there are very few experts and Scotland, in particular, has a globally important liverwort flora. The aim of this work, therefore, is to develop a system for rapid, low-cost identification of liverworts and for the discovery of new species.

Tropical diversity

Our taxonomic research in the tropics concentrates on understanding whole families and genera, especially those that are species-rich yet poorly known and of economic or conservation importance. This research takes a monographic approach – studying a whole family irrespective of its distribution (as opposed to a floristic approach, which studies all the plants within a defined geographic region). RBGE's scientists wish to understand how factors such as climate change, dispersal and geological events have influenced the development of different species of tropical plants and their distribution, as well as the effect of future climate change on tropical plant species.

Four plant families are the current focus of this research: Gesneriaceae (the

Cryptogamic plants

RBGE has a long-standing reputation for working on cryptogams, which make up around 84 per cent of the world's botanical diversity. The cryptogams (which do not reproduce by seeds) include fungi (including lichens), bryophytes (mosses, liverworts and hornworts), pteridophytes (ferns and horsetails) and algae. The main focus at RBGE is on British habitats and species, but some of the work takes place overseas.

The word 'cryptogam' is still used as a term of convenience, although the organisms in question – including algae, diatoms, mosses, liverworts, hornworts, lichens and fungi – are not regarded by scientists as a coherent group and so in strict taxonomical terms the word is considered obsolete. However, the Garden's Cryptogam Group works on all these 'plants', including fungi, which are technically not plants at all as they are now considered to be in a kingdom of their own. Lichens, likewise, are not true plants as they are composite organisms consisting of a symbiotic association of a fungus with either a green alga or a cyanobacterium (a type of bacterium that obtains its energy through photosynthesis).

Scotland is particularly rich in cryptogams, containing 60 per cent of the European species of bryophytes on just 0.75 per cent of Europe's land surface. On a global scale, 10 per cent of the world's lichens grow in Scotland, on only 0.05 per cent of the world's land surface. These plants are therefore of considerable importance within the Scottish context.

There are relatively few cryptogam scientists in the world, yet the importance of these small plants to ecosystems and habitats is enormous. Algae, for instance, carry out almost half of the photosynthesis on Earth – they fix 105 billion metric tonnes of carbon each year, and produce the oxygen in every second breath we take. The important task of describing, listing and cataloguing them is far from complete and so staff at the Garden have been working on floristic and monographic projects, for example on rust fungi, lichens and bryophytes. Staff liaise closely with special interest groups and specialist societies and take part in workshops, excursions and practical identification classes, all with the intention of raising the national capability in identifying these important plants.

left Scotland has many cryptogam-rich habitats, particularly on the west coast, due to high rainfall, clean atmosphere and suitable habitats.

top right *Mylia taylorii*

middle right *Plagiochila asplenioides*

bottom right *Sphagnum lindbergii*

above Members of the Zingiberaceae family (gingers) that are the subject of research at RBGE. **left** This new species of *Amomum*, described in a PhD thesis, forms part of a revision of the 25 species of *Amomum* in Sumatra. **middle** A new species of *Curcuma* to be described in the *Flora of Thailand*. Work on this species is a good example of collaborative research between RBGE, Queen Sirikit Botanical Gardens, Thailand and Singapore Botanic Gardens. **right** Growing in the Wet Tropics House for years, this was identified by a researcher as *Etlingera loerzingii* from Sumatra.

family that includes *Saintpaulia* – African Violet, *Streptocarpus* and *Gloxinia*), Zingiberaceae (the gingers), Begoniaceae (including the genera *Begonia* and *Hillebrandia*) and Sapotaceae (with over one thousand species of mostly tropical trees and shrubs). RBGE's Living Collection is used intensively to support work on the first three families, while work on Sapotaceae is largely field and herbarium based, as living plants would take too much heat and space to cultivate successfully at Edinburgh.

As well as the traditional methods of observation and exploration, the scientists use modern techniques such as phylogenetic systematics – the study of organisms, and their grouping, for the purposes of classification, based on their evolutionary descent. This approach, also known as cladistics, is a special taxonomic technique applied to the study of evolutionary relationships.

RBGE's research in tropical diversity results in the publication of scientific papers and books as well as conservation assessments and online databases. These online resources, such as the Southeast Asian Begonia Database and the Zingiberaceae Resource Centre, have been developed to ensure that the information is widely and freely available to all who need it, including scientists, foresters and planners. These resources are especially valuable in areas that lack well-stocked libraries and have technology tailored to the needs of those

below RBGE's scientists recognise the importance of making their research available to international academic collaborators, planners and land managers.
left The Zingiberaceae Resource Centre is a web-based database which brings together information about species within the family Zingiberaceae into one place, sourced from libraries, herbaria and botanic gardens around the world.
right The Southeast Asian Begonia Database. Its purpose is to stimulate the production of Floras, which are urgently needed for this large and diverse genus, through the provision of regional species lists.

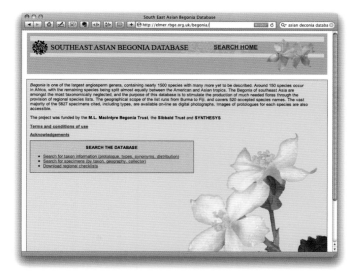

above These three *Begonia* species were photographed on the Indonesian island of Sumatra and represent a few of the 75 species found there. *Begonia gracilicyma* **left** and *B. padangensis* **middle** grow in montane forests in the west, whilst the species on the right from north Sumatra is one of the many new species, yet to be given a name, found during recent expeditions. On Sumatra only about half of the *Begonia* species have been described. Collections of living, DNA and herbarium material brought back to RBGE are transforming our understanding of this incredibly species-rich genus.

right The living Gesneriaceae collection is essential to RBGE's research into particular fields of research, such as evolutionary development and cytology. RBGE staff and a worldwide network of collaborators are using the group as a model to investigate the evolution of floral form and vegetative habit. These are linked to pollinators and the specific ecological requirements of the species, respectively. The African genus *Streptocarpus* exhibits a great morphological diversity exemplified here by the putatively long-tongued fly-pollinated *S. cyaneus* (right), or the unifoliate bird-pollinated *S. dunnii* (far right below). Asiatic Gesneriaceae exhibit an even higher floral diversity and include alpine plants as well, such as *Corallodiscus lanuginosus* (far right above).

who will use them. In this work, RBGE adopts the 'ecosystem approach' to conservation, which is based on the use of scientific techniques appropriate to the scale and complexity of the problem, recognising that humans, with their cultural diversity, are integral components of ecosystems.

Global environmental challenges

Plants are fundamental to human life and are, indeed, the basis for all life on Earth. Without their ability to capture the sun's energy through the process of photosynthesis there would be no life. In maintaining the atmosphere, protecting the topsoil and purifying water, plants provide irreplaceable services to

Plants are the basis of all life on Earth. They provide food, shelter and medicines, and they regulate ecosystem functioning such as the cycling of water and the atmosphere.

above Forests absorb carbon dioxide and release oxygen, they control erosion and store water and they provide shelter and food for numerous animals, birds and insects. Temperate rainforest, Nahuelbuta, Chile.

the ecosystem. For us humans, they provide food, shelter, medicine, fibres and much more. They also enhance our lives though their beauty and are used throughout the world for religious and cultural purposes. Looking to the future, it is plants that hold the genetic stock that will enable us to further improve crops, find new medicine and enhance the quality of our lives.

Yet the threat to global biodiversity is now more severe than at any time in human history with consequences that threaten the functioning of ecosystems and jeopardise basic human needs. There are an estimated 380,000 plant species on Earth and it is possible that two-thirds of these could become extinct by the end of this century.

This would mean the largest extinction crisis in 65 million years, caused mostly by human population growth and changing patterns of consumption. Action is urgently needed to halt the catastrophic decline in the world's plantlife.

In response to this crisis, botanic gardens around the world have developed an ever-increasing focus on conservation over the past decades. Prior to the 1980s, much of their output – such as Floras and monographs – was of conservation use but was primarily produced for academic purposes. The networking organisation Botanic Gardens Conservation International (BGCI) led the change to a more explicit focus on conservation. In response to the World

top right Trees provide timber for building, furniture and firewood. Sawmill near Thimpu, Bhutan.

bottom right Cultivated plants provide an amazing diversity of foods, from onions, potatoes and cabbages, to peppers, courgettes and tomatoes. Market stall, Turkey.

Conservation Strategy, which was published in 1980, BGCI published the Botanic Gardens Conservation Strategy in 1989. This emphasised that botanic gardens had the staff skills and physical resources needed for plant conservation and highlighted important activities that were not being undertaken by other agencies. As a result plant conservation itself has evolved from an emphasis on holding individual plant species in collections to the current goal of protecting plant biodiversity in the widest sense, both in collections and in the wild.

Today the world's major botanic gardens are among the leading agencies working for plant conservation worldwide. They have a vital role to play in conservation because of their traditional strengths in systematic research, public education and practical horticultural know-how. In recent years, botanic gardens have contributed to the formulation of legislation to protect threatened species and habitats and to discourage trade in endangered species. They contribute to the monitoring of plant populations and take practical action for conservation, building effective relationships between land managers, academics and plant conservation communities.

At RBGE, our conservation work has gained momentum ever since the 1980s, when the Garden established a number of conservation initiatives, including the Scottish Rare Plants Project and the

BGCI
Plants for the Planet

above Botanic Gardens Conservation International (BGCI) was founded in 1987, to coordinate conservation action in botanic gardens. Since that time it has held conferences, stimulated regional networks of botanic gardens and worked at a political level to develop conservation policies, agendas and conventions.

International Conifer Conservation Programme. RBGE also contributed to the Government-led process to formulate Biodiversity Action Plans, first for the UK and then for Scotland. In recent years, our research has been closely aligned with the Global Strategy for Plant Conservation, with its aims to understand, document, conserve and raise awareness of plant diversity.

The Garden's research is now focused on seeking to understand critical conservation issues: the reasons for biodiversity loss; climate change; the sustainable use of plants; and the role of plants as contributors to ecosystem services. New species are being described with greater urgency and conservation assessments of threatened species are now a standard procedure in fieldwork. However, these activities cannot realise their full potential unless the results are effectively communicated and so RBGE has promoted all forms of communication with particular vigour, from traditional

journal publications to specialist databases to the creation of the John Hope Gateway, the main purpose of which is to explore conservation issues and solutions. Collectively these activities are aimed at gaining and sharing a better understanding of the world of plants and their essential contribution to our world.

Practical conservation

There are two aspects to the challenge of minimising biodiversity loss. The first is to provide conservation assessments for species or regions which are lacking in data, while the second is to use these assessments to develop conservation strategies to protect threatened species. For this to happen, RBGE is working to improve the link between the efforts of the taxonomic community to describe biodiversity and the efforts of the conservation community to protect it. This information flow is urgently required to reduce the large backlog of species whose

The Global Strategy for Plant Conservation

The Global Strategy for Plant Conservation was adopted by the Conference of the Parties (COP) to the Convention on Biological Diversity at a meeting in the Netherlands in April 2002. A strategy designed especially for plants, to work within the boundaries of the CBD, was first mooted at the International Botanic Congress (IBC) held in Missouri in 1999.

The Strategy is arranged in 16 outcome orientated targets which were to be reached by 2010 with the ultimate objective of halting the current and continuing loss of plant diversity. It was intended to be a tool to enhance the ecosystem approach for the conservation and sustainable use of biodiversity and focus on the vital role of plants in the structure and functioning of ecological systems. There are defined objectives subdivided into targets: understanding and documenting plant diversity; conserving plant diversity; using plant diversity sustainably; promoting

education and awareness about plant diversity; and building capacity for plant conservation.

RBGE is contributing to at least 8 targets including:
1. A widely accessible working list of known plant species, as a step towards a complete world flora
2. A preliminary assessment of the conservation status of all known plant species, at national, regional and international levels
3. Development of models with protocols for plant conservation and sustainable use, based on research and practical experience
7. 60 per cent of the world's threatened species conserved *in situ*
8. 60 per cent of threatened species in accessible *ex situ* collections and 10 per cent included in recovery programmes (bottom right: the Arran whitebeam (*Sorbus arranensis*) growing *ex situ* at RBGE)
14. The importance of plant diversity and the need for its

conservation incorporated into communication, education and public-awareness programmes
15. The number of trained people working with appropriate facilities in plant conservation increased, according to national needs, to achieve the targets of the Strategy (top right: Training workshop at the Royal Botanic Garden Serbithang, Bhutan)
16. Networks for plant conservation activities established or strengthened at national, regional and international levels (middle right: A PlantNetwork meeting, Benmore Botanic Garden, Scotland. PlantNetwork is the plant collections network of Britain and Ireland)

In October 2010 the COP to the CBD agreed to continue with the GSPC for a further ten years, to 2020. The objectives of the 16 targets remain unchanged but many of the percentages have been increased and clarified.

---

conservation status is currently unknown.

As our botanists are frequently working in the field, they have the practical knowledge and insights into local problems that may be causing a species' decline. Therefore they are in an ideal position to provide the necessary data to those engaged in conservation planning. RBGE's botanists make an important practical contribution to conservation by producing the prerequisite material: field guides, regional check lists, inventories and habitat assessments. Conservation strategies are now focused not just on protecting sites where threatened plants are represented, but on a dynamic approach which aims to conserve evolutionary and ecological processes such as the development of networks of protected areas.

This strategy is especially important where the plants involved are considered 'taxonomically complex' because the individual species are not distinct. For example,

in the genus *Euphrasia* (the eyebrights), RBGE carried out detailed studies to assess the plants' status for conservation purposes, but the species defied simple classification into discrete entities. This meant that a conventional species-based framework for conservation would be inadequate.

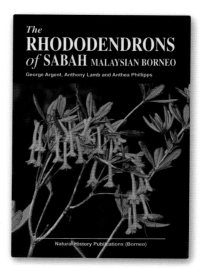

The International Conifer Conservation Programme

The International Conifer Conservation Programme (ICCP) was launched at RBGE in 1991 to help address the serious problem of species and habitat loss and to raise public awareness of conifers and their importance. The Programme combines research in conifer taxonomy with conservation, population genetics, cultivation and international capacity building.

Many of the world's 722 naturally occurring conifer taxa come from temperate rainforests, which are very limited in their extent and distribution, especially compared with tropical rainforest. The largest remaining regions of virgin temperate rainforest lie within the southern hemisphere and include areas of New Zealand, Tasmania, Chile and Argentina. In these countries conifers are often confined to the mountains, where they form the dominant component of the vegetation. Today, because of heavy logging and dramatic changes to land use, many of these species, and the forests where they grow, are facing unprecedented destruction. Numerous factors have contributed to the demise of conifers in the wild, including overgrazing, forest fires, volcanic activity, open-cast mining and flooding. Drought caused by human-induced changes to the water table and out-pollination with exotic commercial species all contribute in their own way to the demise of conifers in the wild.

The first phase of the Programme was to set up a network of 'safe sites' throughout Britain and Ireland to accommodate breeding populations of threatened conifers. With their diverse climates and soil types, these countries are ideally placed to accommodate conservation collections of conifers. Scotland, in particular, has a long and success- ful history of growing conifer species outside the countries where they naturally occur. The ICCP has now established a 'safe site' network of over 120 properties.

The intention of this network is that each threatened species would be represented by carefully structured populations using seeds collected from throughout their remaining geographic ranges, creating a 'living gene bank' for the species. Today many of these breeding populations are being used for research relating to conifer taxonomy and conservation genetics. The long-term aim is that these populations can be used to help replenish genetically depleted wild populations, should this be necessary.

Systematic research on conifers at RBGE has fully utilised this known wild origin living material, hence the Living Collection has benefited greatly from the ICCP, and now contains more than 4,000 accessions of conifers, representing 467 species and 634 taxa within the four Gardens and the external network of 'safe sites'.

There is still an urgent need for the countries where threatened conifers occur in the wild to establish their own conservation measures, and the ICCP has widened its remit to support this through activities such as field surveys, conserva- tion assessments and especially international capacity building. The ICCP works with many botanic gardens, NGOS, national parks and agencies in countries including Chile, Cuba, Laos, Mexico, New Caledonia, the Solomon Islands, Turkey and Vietnam. Higher-level training, both at home and abroad, has also been a very important part of the work and the ICCP encourages students from these countries to study at RBGE.

top Monkey puzzle (*Araucaria araucana*), in Conguillio National Park, Chile, one of seven threatened species of conifer in Chile.

middle Devastated forest in central Chile, deliberately destroyed to create arable farmland.

bottom left Propagation workshop, Valdivia region, Chile.

bottom right Collecting seed of the Serbian spruce (*Picea omorika*).

Instead, management measures need to be targeted to conserve sites where 'hotspots' of *Euphrasia* biodiversity occur, preserving the evolutionary processes of hybridisation and natural selection that generate this very biodiversity.

RBGE also carries out a significant amount of specific conservation research, which goes beyond taxonomy. Such work includes projects focused on declining populations or threatened species, with studies into different aspects of the plants involved, including life cycles, pollination, population dynamics and genetics, as well as the relationship between species and habitat. This multidisciplinary approach was recently used in a study of sub-arctic willows. Another example was a recent study on the threat of hybridisation of the native bluebell (*Hyacinthoides non-scripta*) with the Spanish bluebell (*H. hispanica*) and hybrids (*H. x massartiana*) and was commissioned to better understand the process and to find possible solutions.

The presumption is that *H. hispanica*, which was introduced to the UK in the 17th century, naturalised in gardens and subsequently escaped from cultivation. It seems that the main threat to *H. non-scripta* comes from increasing hybridisation and from losing ground, literally, to a superior competitor. However, little data has been available on the numbers of non-native bluebells, the extent of their co-occurrence with natives and characteristics such as dispersal, fertility and responses to climate. This information is vital for worthwhile conservation measures to be put in place, and so RBGE has been studying natives and non-natives in the wild, in the nursery trade and in common gardens across southern Scotland in an attempt to answer these questions.

The writing of inventories – lists of plants present in a defined region at one period of time – has long been a traditional role of botanic gardens as part of ongoing taxonomic research. The gardens now are

Taxonomically complex species may be very difficult to distinguish, making conservation action difficult. This was the case with Scottish species of eyebright (*Euphrasia*) which were highlighted in the *UK Biodiversity Action Plan* as a group that required taxonomic determination before conservation work could start. This work fell to RBGE's scientists.

top left *Euphrasia officinalis* agg.

top right Typical *Euphrasia* habitat, Strathy Point, Sutherland, Scotland.

able to offer their expertise to agencies who are undertaking inventories for conservation purposes. Staff from botanic gardens often act as consultants, assessing critical species or plants that are difficult to identify. RBGE is currently undertaking inventory research in threatened habitats within Asia, Africa and Latin America.

Many conservation projects require monitoring, which is the recording of individual species of habitats over a defined period. This recording is normally undertaken over several years in order to establish changes in numbers within populations. Monitoring is often done to assess the impact of an activity on a particular plant species. For example, the Scottish Rare Plants Project is monitoring a number of species to assess their status and, if they are found to be declining, to understand the cause. RBGE is also establishing monitoring sites high in the Scottish mountains for alpine plant species and snow-bed mosses and lichens. These are used to help understand how climate change will impact on vegetation.

In the field or in the Garden

One of the major issues in plant conservation today is the relationship and the balance between '*in situ*' and '*ex situ*' conservation. *In situ* means the conservation of plants within their natural habitat, while *ex situ* refers to the conservation of plants outside their natural habitat – such as in specialist collections and botanic gardens.

In situ conservation is usually considered to be the most desirable approach for conserving both habitats and species. It can be an efficient and effective approach to maintaining biological diversity through the management of wild land and the creation of protected areas. However, it was once assumed that establishing national parks, nature reserves and other means of protecting large tracts of land would safeguard species and habitats but now there is serious concern that the fragmentation of the landscape and distances between protected areas may preclude natural movement of species. For example, species that may have to colonise northwards in the northern hemisphere under the pressure of global warming may

195

A Partnership for Plants in Scotland

Launched in December 1991 (initially as the Scottish Rare Plants Project), this initiative is funded by the Government conservation agency Scottish Natural Heritage and located and managed at RBGE. The initial objectives of the project were to compile dossiers of threatened species, monitor populations, support the recovery of rare populations and provide advice and training to other conservation bodies. During the first phase an enormous amount of fieldwork was undertaken to record the status of threatened species, including *Lychnis alpina*, *Polygonatum verticillatum* and *Oxytropis campestris*. Reintroduction programmes were established for the sticky catchfly (*Lychnis viscaria*) and oblong woodsia (*Woodsia ilvensis*).

In the current phase of the project there is a greater emphasis on education and partnership working. The key aim is to raise enthusiasm for Scotland's plantlife, harnessing the energy of Scotland's people to acquire a greater knowledge of our flora and to help people develop new skills and then apply them to the conservation of plants throughout Scotland. Workshops with special interest groups are held to help provide training in species identification and the project hosts public lectures at the Garden and other centres. This project is now well established as a long-term commitment and plans are being developed for further work and even greater partnership working, both within the Garden and beyond.

Reinforcement of Lychnis viscaria

A survey of Scottish localities in 1991–1993 recorded that several populations were declining, with the main threat being shading from surrounding vegetation. The oldest population recorded in Britain was known to be in Holyrood Park in the centre of Edinburgh. By the early 1990s, this had been reduced from a thriving population to just four clumps, but in this case the threat to the remaining plants was from fire. Given that this was the last extant colony in the Lothians, a reinforcement programme was instigated in conjunction with Historic Scotland, the owners of the site, and in consultation with Scottish Natural Heritage. Plants were raised at RBGE from seed collected in 1993 and 20 individuals were planted at a location in the vicinity of, but separate from, the remaining native plants.

During the year after planting, 80 per cent of the plants were recorded as having established, despite a very dry summer. The exact location of each plant was recorded and monitoring has been undertaken at various intervals since planting. To date, all those that established have gone on to survive but few, if any, seedlings have established. Monitoring will continue but the project will not be deemed to be a success until new plants develop, flower and set seed themselves, thereby creating a new, vibrant population.

Reintroduction of Woodsia ilvensis

The oblong woodsia *Woodsia ilvensis* is a small fern that is relatively abundant in northern Europe, the north-western oceanic margin of its geographic distribution. In Britain, the mid 19th century fashion for fern collecting drastically reduced populations of many fern species, including *W. ilvensis*.

Following RBGE research in 1995, it was established that there were no more than a hundred clumps left in the wild in Britain, and that these were distributed among six broad localities: three in Scotland, one in England and two in Wales. The next step was to establish *ex situ* conservation collections of *W. ilvensis* in order to halt any further loss of British genotypes and to provide material for possible reintroduction. Intensive horticultural trials at the Garden resulted in the successful cultivation of the species. By 1998, a conservation collection had been formed of about 2,500 young plants growing in pots outdoors, representing all extant British populations.

On a national level, the catastrophic decline of *Woodsia ilvensis* in Britain resulted in the species being included in various conservation

not be able to 'jump' sufficiently far to reach the next suitable site.

Where the number of plants of a certain species has become insufficient to maintain genetic stocks, *in situ* conservation is no longer possible and *ex situ* conservation is used to complement this. It is implemented to provide 'reservoir populations', or stocks, which can be used to support the survival of species in the wild by reintroduction or restocking, or to support habitat restoration

and rehabilitation. *Ex situ* conservation acts as an insurance policy by holding stocks in long-term storage for future needs. This work has benefited from modern molecular research, especially regarding sampling techniques, the number of populations to sample from and representative numbers of individual plants to hold. *Ex situ* collections also provide a valuable resource for research into basic aspects of species biology, which may be critical in

left Oblong woodsia (*Woodsia ilvensis*) growing wild in Wasdale.

above An RBGE conservation scientist studies a successfully reintroduced sticky catchfly (*Lychnis viscaria*) on a hill in Edinburgh.

right Sticky catchfly (*Lychnis viscaria*) was the first plant restoration project undertaken by RBGE.

priority lists. In 1998, a Species Action Plan recommended positive intervention to restore populations by translocation of the species. RBGE was in a good position to carry out this restoration thanks to its research and the numerous young plants from each population already in cultivation. These were planted out at sites in Teesdale and in the Southern Uplands in 1999. Monitoring took place at all three sites to check how many survived, and of these how many were shedding spores. Like the *Lychnis viscaria* reintroduction trials, it will be many years before the project can be assessed as being successful.

devising appropriate and effective *in situ* programmes.

Ex situ conservation has its limitations as it can be very difficult to obtain a representative sample of a species to hold in a collection, and some species are technically difficult or costly to maintain in this way. Co-evolution with other organisms such as pollinators and seed dispersers cannot continue in *ex situ* collections, and certain interactions such as competition or population dynamics may be more difficult to study in an *ex situ* environment. There is the added risk of some authorities considering it a substitute for *in situ* conservation and thereby putting remaining natural areas at risk. For these reasons the most satisfactory way to proceed with conservation projects usually requires an integrated strategy, combining *in situ*, *ex situ* and other relevant approaches. Virtually all of RBGE's current conservation projects are integrated and are undertaken

above *In situ* and *ex situ* conservation techniques involve the conservation of plants in their natural habitat and away from their natural habitat (for instance within botanic gardens) respectively. Ideally the two should be practised together, along with other approaches such as legal protection, research and public education.

top row *Ex situ* conservation techniques include the long-term storage of spores and seeds in seed banks (**left**), cultivation of living plants to study their propagation and growing requirements (**middle**) and the collection of seeds from the wild to bring into protective custody.

bottom row *In situ* conservation techniques include species assessments (**left**), plant restoration, back into the wild (**middle**) and habitat surveys (**right**).

in partnership with other organisations that can provide complementary skills, resources and connections with the country involved.

RBGE views both *in situ* and *ex situ* conservation as vital parts of its work in protecting plantlife. Our staff are involved with *in situ* projects throughout the world, including projects in Chile, China, New Caledonia and the Congo, as well as at home in Scotland, where they carry out activities such as site selection, threatened species mapping, scientific management, research and biodiversity surveys.

In caring for our extensive *ex situ* collections, we ensure that these are properly targeted for conservation use. Just because a botanic garden holds threatened plants, it does not necessarily mean that the specimens are of any value for conservation. A garden's *ad hoc* conservation collection includes those threatened plants that have been amassed over time for education or display rather than conservation. The plants may have poor records with the important information such as collection

location or habitat unknown. Targeted *ex situ* conservation collections, on the other hand, are characterised by detailed plant records, significant numbers, a good genetic representation of particular populations and being used actively for research or conservation, with the country of origin involved in some way, such as through a conservation partnership.

Gene banks, such as seed banks and spore banks, are a particular type of *ex situ* conservation concerned with the long-term storage of plant parts for conservation, breeding and research purposes. The facility for the UK is the Millennium Seed Bank at the Royal Botanic Garden, Kew's Wakehurst Place in Sussex. With this large, well-equipped seed bank relatively close by there is no need to duplicate this work, and so there is no long-term seed bank at RBGE.

Most seeds can be preserved for many years when properly stored, and so the aim of seed storage is that species can be re-established in the wild using seeds from the bank when conditions become suitable.

opposite top Seed banks provide an efficient way of storing large numbers of seeds over a long period of time in a relatively small space. RBGE does not have a large seed bank for long-term conservation but stores seeds for short periods of 5–20 years as part of the ongoing management of the Living Collection.

top Spores being cleaned and prepared for storage.

bottom Seeds stored in jars in a freezer.

While they are an important and efficient component of plant conservation, seed banks can never substitute for conserving whole plants and the knowledge and skills gained by horticulturists growing the plants concerned.

Return to the wild

The replacement of plant material into the wild is often considered the ultimate goal of *ex situ* conservation, but for collections to be of real use for this purpose they have to be genetically representative and must consist of enough individual specimens. When properly executed and with the necessary scientific back-up, reintroductions are one of the most exciting activities that botanic gardens are currently engaged in, combining the resources and expertise of scientific and horticultural staff.

As some plant species have become increasingly scarce and, indeed, as individual populations have been lost altogether, it is becoming ever more important to assess the possibilities and techniques of returning plants to the wild. The reintroduction of

plants into the wild raises all sorts of logistic, scientific and moral issues, and there are some variations in the methods used:

Augmentation is the process of propagating plants from an existing population and returning them to the same population. The purpose is to increase a population with material originating from the same genetic pool, and the process may be used when the number of individual plants has become so low that the continued existence of the population is threatened. Where reproduction is impaired by a small genetic pool, an augmentation project may also involve introducing propagules from another population.

The process of collecting plant material in the form of seeds or cuttings from one or more still surviving populations, growing them into plants and then introducing them back into sites where the species once occurred is called *reintroduction*. It is used to restore or increase the number of populations. This specific use of the term 'reintroduction' should not be confused with

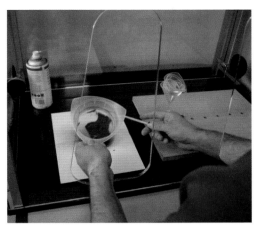

Phenology

Phenology is the study of seasonal phenomena. For plants, this usually relates to events such as bud burst, flowering, fruiting and leaf drop. It is a topical subject for study as it is probable that the increasingly early dates for these events provide evidence for climate change. Weather can change quite dramatically from year to year and so only long-term studies will reveal long-term trends.

Phenological records can be useful in many branches of research – some use this data to investigate the correlation with climate patterns, while others look more deeply into the relationship between the patterns and the actual physiological triggers for the events, such as whether a particularly warm summer and autumn would increase the number of flowers in the following year.

The earliest phenological data recorded from RBGE at its present site is the first-flowering dates of 24 plants recorded in 1850 by then

Curator James McNab. He continued recording the first-flowering dates of some 90 species until 1878. Thereafter the work was continued by the next two Curators, John Sadler until 1882 and then Robert Lindsay until 1895, when recording may have ceased. Monitoring was continued in the 20th century and records exist in the RBGE archives for observations between 1906 and 1939. Unfortunately, no records can be found for the period 1940–2000.

In 2002, recording resumed under dedicated volunteers, supported by RBGE staff. The project involves daily and weekly monitoring, recording the frequency and duration of flowering for over 100 species and noting the stages with each flowering season and the changes in foliage. This data will increase the understanding of the mechanism in plants which responds to climate changes and will enable scientists to predict how plants will respond to climate change.

the commonly used generic meaning of all forms of replacing plant material in the wild, sometimes also called 'recovery' or 'restoration'. When plants grown from collected material are introduced to new sites, then the process is called *introduction*.

Mitigation is a term primarily used for moving plants to a suitable location from a site that is due to be cleared for construction. The process includes augmentation, reintroduction and, most frequently, introduction with the intention of reducing the adverse impact of development projects. Mitigation is not a means of enhancing the viability of rare plant populations.

Like mitigation, *plant rescue* involves the removal of plants shortly before their otherwise impending destruction. It usually involves storing plants at a temporary site until a suitable plan can be developed for their long-term survival.

A concern about the practice of augmentation, mitigation and introduction held by many conservationists is that the perceived ease of moving plants from a potential development site can be used as a justification to develop land that contains threatened or important species and habitats. However, reintroduction projects are time-consuming and can be logistically complicated. It is essential that detailed records are kept of the plant material with exact references to its original provenance and the planting locations. For many years after the original replacement, regular assessments are required in order to record the longevity of the mother plants and the occurrence of new seedlings. As most projects have only been undertaken in the last 10 to 15 years, the long-term benefits are still to be assessed.

Horticulture – growing expertise

The expertise of horticultural staff is essential for many of the conservation techniques described in this section and horticulturists within botanic gardens are especially suited for conservation work. Their skills are markedly different from those required by either commercial or amenity horticulture. Botanic garden horticulturists have the know-how to germinate a great diversity of seed, including those with special dormancy-breaking requirements, and the ability to cultivate an extraordinarily wide range of species, many of which have seldom been cultivated before and which are often far more demanding than commercially available cultivars. Staff from botanic gardens also understand the necessity of keeping accurate written records while carrying out normal horticultural operations.

Horticulturists have an important role to play in whole habitat restoration, a concept that is becoming more widely considered as the relentless loss of natural habitats continues. It is quite impossible to recreate exactly a wild habitat, but research in this area is increasing and many large-scale field trials are now taking place. Though it may take many centuries for a new habitat to fully re-establish the complex interactions between plants, animals and mycorrhizal organisms, efforts are being made to recreate habitats that have been lost. This work requires both the expertise of ecologists with their understanding of whole habitats and the skills of horticulturists who understand the cultivation of the individual species involved.

Botanic garden horticulturists have always held a wealth of knowledge about

the cultivation of plants, but traditionally there was no platform to publish such information. *Sibbaldia: The Journal of Botanic Garden Horticulture* was founded by RBGE in 2003 to provide an opportunity for horticulturists to record and share this valuable knowledge. Horticulturists are now recording the results of their work in more detail, so that the necessary information and techniques will be available when required.

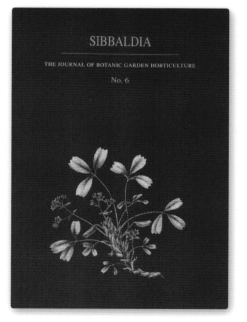

above Horticultural expertise is increasingly being valued in conservation and restoration projects. While scientific knowledge, legal protection and habitat management techniques are vital, horticultural expertise also has an important role to play. Horticulturists have the skills to germinate, propagate, cultivate, record and reintroduce plants. Germinating fern spores and seeds of *Streptocarpus* in the growth rooms for research.

left Communicating successes and techniques – *Sibbaldia: The Journal of Botanic Garden Horticulture* was launched in 2003 to encourage horticulturists to record and share their knowledge – knowledge that is often assumed to be known but is seldom committed to paper.

Partnerships 'for a better future'

In the years ahead, there will be an increasing demand for botanic gardens to do more restoration work, more research and more public education. The climate is changing, the human population is growing faster than at any time in history and people are demanding higher standards of living than ever before. This means that wild land continues to be cleared for agriculture and urban development, resulting in further reductions in natural habitats and the loss of species. As Scotland's leading international player in plant biodiversity, there is no doubt that RBGE will devote ever more energy and resources towards plant conservation.

We will continue to build partnerships with organisations that work to improve people's livelihoods through the sustainable use of plant species. Many people in developing countries depend on their local plants but burgeoning international markets are driving excessive gathering, which puts pressure on wild populations. Plant resources cannot be renewed if they are over-exploited. Our role is to help develop strategies for the sustainable use of plants, increasing our collection and documentation of useful plant species and targeting our taxonomic research on plant groups with economic uses. We work in collaboration with others, including foresters, ethnobotanists and social anthropologists, to develop means by which plants can be harvested without threatening their long-term survival.

RBGE's scientists are also looking to the bigger picture of the role plants play in the functioning of the planet. The ability of the Earth's ecosystems to sustain future generations can no longer be taken for granted. 'Ecosystem services' represent the planet's life support systems, including nutrient cycling, soil production and stabilisation, clean water provision and climate regulation. The last 50 years have witnessed an unprecedented degradation of natural ecosystems. This has led to a substantial loss of species and to the degradation or loss of the services ecosystems provide us with, such as clean water and food.

The challenge is to slow down or reverse the degradation of natural ecosystems, but to do this we need a more complete understanding of how plant groups contribute to the maintenance of key ecosystem services and the importance of biodiversity for

below left In Scotland with Scottish Natural Heritage and landowners as part of the Partnership for Plants in Scotland. Staff collecting cuttings for the Target 8 project of the Global Strategy for Plant Conservation.

below right In Belize, at Pine Ridge, in partnership with the NGO Programme for Belize and the Belize Government. Staff and Government officials monitoring populations of *Pinus caribaea* at the Rio Bravo Conservation and Management Area.

right In Peru in partnership with the NGO Asociación Peruana para la Promoción del Desarrollo Sostenible (APRODES). RBGE staff and Peruvian collaborators in a Darwin Initiative project inspect *Podocarpus* seedlings in a nursery.

below left In Soqotra giving training to staff at Adeeb's nursery, Hadibo. The nursery was a partner organisation with RBGE during a Darwin Initiative project on Soqotra.

below right In Lao PDR, RBGE's Head of Education, Leigh Morris, teaching the Certificate in Practical Horticulture at the new Pha Tad Ke Botanical Garden along with students from Luang Prabang Agricultural College.

maintaining various ecosystem functions. This means closer collaboration with Earth systems scientists, a new venture for RBGE but a logical extension of our long-standing work in building partnerships with other institutions and disciplines. We will also be working with social scientists to enable our research programmes to address the links between biodiversity, ecosystem services and society, and how human societies can adapt to environmental change. RBGE has a world-class reputation in inventory and systematic research in many of the regions and habitats that have been identified as critical for the world's ecosystem services. Thus our core work, and cumulative expertise, has a vital new application in helping to understand our planet and to conserve its diverse and invaluable plantlife.

Enlighten, inspire and inform

EDUCATION AND PUBLIC ENGAGEMENT AT THE GARDEN

The Garden has always been involved with education, from its very foundations in the 17th century when it was created to teach medical students and apothecaries to identify plants correctly and distinguish their medicinal properties. As botany became a subject in its own right, botanic gardens looked beyond just the medicinal use of plants and taught wider aspects of plant science, from plant physiology and classification in the 18th century to the cellular and molecular approaches of the 20th century. In the 21st century, with the world's plantlife facing unprecedented threats, RBGE's educational work has a vital role to play in plant conservation.

left The glasshouses offer an inspiring living environment for a lesson in human ecology: Edinburgh schoolchildren practise shooting through a blow pipe as they learn about life in the rainforest.

The Living Collection is at the heart of all education at RBGE. There are now over 13,500 species of plants growing in the four Gardens, and to some extent every plant has an educational role and a story to tell. RBGE's Living Collection contains an astonishing diversity of plant forms and represents many climates and habitats. It enables school children to explore the plantlife of rainforests, deserts and mountain peaks and PhD students to analyse the molecular differences between species or between different populations of the same species. It invites all who come – whether on a formal education course or a recreational visit – to pause for a moment and gain a greater awareness of the beauty and importance of the world's plantlife.

With the sheer range and accessibility of its plant collections, the expertise of its staff and the outstanding resources of the Herbarium, Library and Archive, RBGE has a unique capacity to provide education for every age and stage, from nursery teacher training, through a comprehensive schools programme to all stages of higher education and a diverse spectrum of adult education. This educational role reaches all four Gardens and many projects around the world.

Working with schools

At every level of education, and whatever the age of the pupil, the fundamental mission is the same: stimulating awareness and appreciation of plants through practical engagement with the real thing. This begins with children from the earliest primary years who come with their teachers and classes to learn about how plants live and how much we need them.

RBGE has long been at the forefront of

pioneering teaching methods and today the children learn through dynamic and interactive styles of education. They are encouraged to touch plants, to grow them, sometimes to taste them and always to watch them closely – observation is a key skill of a scientist and one that can be honed at a young age. The programmes are devised by full-time teaching staff working with the Garden's scientists and are closely linked to the national curriculum. This work also fulfils the Government's wider educational objectives, such as encouraging healthy living and eating, raising environmental awareness and outdoor learning. Most programmes are cross-curricular, for example, classes for secondary schools are focused on learning requirements for biology, geology or art, but they combine elements of all three.

The education staff are based in the Edinburgh Garden but their work reaches far beyond the capital city. School groups come on day trips from as far as Aberdeen, Ayr and Newcastle, and special events draw pupils from even further afield. Educational events are expanding at all the Regional Gardens, and each autumn, the Edinburgh-based team spends a week at Benmore Botanic Garden, offering activities for schools throughout the area. The education staff also participate in RBGE's conservation projects around the world, sharing their skills with staff in developing botanic gardens.

However, staff and time resources are limited and so to reach more children, the education team works to transfer their skills to schools through teacher training. A lively programme of Continuing Professional Development (CPD) enables teachers to come to the Garden and learn new ways to teach through plants, whether gardening,

below RBGE's education programmes now reach from pre-school to PhD. Nursery children explore the world of plants through imaginative, fun-filled activities such as story telling in the yurt, sensory exploration and teddy bears' picnics.

storytelling, arts and crafts or science skills. Like their classes, the teachers often come for return visits and so the education team constantly devises new ideas and programmes.

Teachers and children can learn together through RBGE's Children's Garden, a small plot behind the Fletcher Building where children can gain hands-on experience of looking after produce throughout an entire growing season, from planting seeds to harvesting the fruit and vegetables. For many years this was run as an after-school club with a small number of children each year, but the Garden is now working with a local primary school to engage more pupils and teachers in learning the skills to develop their own garden in the school grounds. The Children's Garden provides a valuable learning experience for all, as the children teach each other and work together, tending their plots and tasting their own produce at the end-of-term barbecue.

Higher education

School leavers have the opportunity to study at the Garden with every step of higher education now available, from HND/BSC to MSC and on to PhD and beyond. This makes a vital contribution to botanical and horti-cultural training in the UK as opportunities to study plants at a higher level are declining elsewhere. At many universities, the botany courses have now been amalgamated into general biology degrees and the focus of research is often at a molecular level rather than the whole plants, with very few univer-sities retaining their own plant collections. There is no shortage of interested students, however, and all of the Garden's higher education courses are in great demand. Students from throughout the UK and around the world are attracted by RBGE's unparalleled learning environment, where the traditional teaching tools of labs, library and lecture rooms complement the vast, colourful classroom of the Living Collection

right Young trainees proudly tend their produce in RBGE's Children's Garden, where pupils and teachers learn the neces-sary skills to develop their own plots back in their school grounds. So much can be taught through gardening: counting seeds and measuring space, teamwork and planning, organic cultivation and composting, as well as the pleasure of eating home-grown food at the harvest time barbecue!

above Specialist teaching staff devise RBGE's programmes for secondary school pupils, augmenting the national curriculum with the Garden's own blend of hands-on, experiential learning.

outdoors and under glass and the expertise of the RBGE botanists and horticulturists.

All higher education courses are externally validated and run in partnership with a range of universities and colleges. RBGE has particularly close links to the University of Edinburgh, with a long-standing relationship throughout the Garden's history. The HND/BSc course in Horticulture with Plantsmanship is a collaboration with the Scottish Agricultural College and Glasgow University. It integrates the practical experience of plant cultivation and management with advanced knowledge of plant distribution, classification and identification. The course evolved from the long-running Diploma in Horticulture, Edinburgh (DHE), which launched the careers of many of the country's top horticulturists. The School of Horticulture can trace its roots back more than 100 years – it was initially aimed at foresters and plantation growers for the

British Empire, before turning its attention to Parks Department horticulture in the 1950s and then on to its current focus of botanic garden horticulture from the 1990s.

The next step on the education ladder is the Garden's renowned MSc course, 'The Biodiversity and Taxonomy of Plants', which was established in 1992 with the University of Edinburgh in response to growing demand worldwide for scientists with the skills to identify, classify and study plants. It equips students for careers in plant science such as conservation work, taxonomic research and collection curation, as well as further study at PhD level. Practical assignments and fieldwork are an integral part of the course, including a two-week field trip to Belize to study tropical plants and learn ecological survey techniques.

All PhDs are offered as jointly supervised projects involving RBGE and a lecturer from a collaborating university. The diverse

The rich resources of the Garden create an ideal setting to study 'Plantsmanship'. The School of Horticulture has been in existence for more than 100 years and the HND/BSC course in Horticulture with Plantsmanship combines practical experience of cultivation with advanced knowledge of plant distribution, diversity and identification.

left Three images from education classes at RBGE the 1970s.

right Students studying plants in the wild and gaining practical knowledge from garden staff.

expertise of RBGE's staff ensures that the Garden can offer supervision for all types of systematic research, from traditional taxonomy, where a student might revise a genus, to the most cutting-edge molecular or evolution research. The use made of the Garden's Collections is as diverse as the subjects of research. Some may only require the Herbarium and Library, while others may need to use the Living Collection to observe and record a plant's morphology, extract DNA or grow plants to become familiar with cultivation techniques.

RBGE also makes a wider contribution to plant biology teaching at university level, with many staff giving lectures at the University of Edinburgh in subjects such as plant diversity, geography and evolution. All Scottish universities offering botany or plant science visit the Garden for specialist tours and courses, mostly day-long visits, incorporating an introductory lecture about

the work of the Garden and visits to the Herbarium and Living Collections to see the diversity of plant forms and functions. The Garden also has a close relationship with Edinburgh College of Art, offering informal collaboration for student projects as well as lectures and regular Garden visits for students of Landscape Architecture. Since the foundation of the course in the 1970s, Landscape Architecture students have visited the Garden for one day per week during the first two years, studying subjects such as soil science, ecology, arboriculture and plant identification, with at least one session per week spent sketching and studying plants from the Collection.

Adult education

RBGE also offers an ever-expanding programme of adult, or continuing, education, covering horticulture, botany and environmental subjects, as well as art, photography

and well-being. There is a rich variety of courses available, some long-term and some short, one-off courses, often taught in the evening or at weekends. From lichen identification to birch craft, or digital photography to Chinese brush painting, all courses make extensive use of the Living Collection and all embody RBGE's educational mission: getting people close to and appreciating plants.

For many years, the Garden has taught the Royal Horticultural Society (Level 2) Certificate in Horticulture, which provides a broad-based theoretical understanding of horticultural techniques and plant biology. RBGE has recently advanced and consolidated its expertise in hands-on teaching with the launch of its own certificate-level courses in Practical Horticulture and Practical Field Botany. Garden staff have always taught these skills, but devised the course in response to demand for a formally structured course with a qualification. Both practical courses have been integrated into RBGE's degree programmes so that all students leave with competence in these core skills. The Certificate in Practical Horticulture has been adopted by the Cornwall-based Eden Project and is now developed and marketed as a partnership between the two organisations.

above Training the botanists of the future through the Garden's renowned MSc in the Biodiversity and Taxonomy of Plants (top right), which attracts students from around the world. Many stay on for further study here at PhD level (left, upper and lower), using the Garden's resources including the Living Collection for in-depth research into subjects such as plant systematics and evolution.

left and opposite From Chinese papermaking and botanical art to moss identification and digital photography, the wide spectrum of arts, crafts and science courses in the Adult Education Programme aims to entice all abilities to explore the world of plants through diverse media.

Because the fundamental skills involved in plant identification and care are the same all over the world, these qualifications can be taken wherever RBGE is working. The Practical Horticulture course has already been taught at projects supporting new botanic gardens in Laos, Oman, Soqotra and Turkey.

The provision of certified, professional-level short courses has expanded to include RBGE Diploma programmes in Botanical Art, Garden Design and Herbology – the study of the traditional uses, cultivation and conservation of herbs. Like the Practical Certificates in Horticulture and Field Botany, these courses have proved extremely popular, both with expert amateurs and those seeking the first step to a change of career. They represent an important development in RBGE education, a new way in which we can share our expertise and enable others to explore the world of plants.

Capacity building

The Garden is involved in partnership projects in more than 40 countries, and many of these involve capacity building and training. The examples selected here are those based on helping the partner countries to develop new botanic gardens and the horticultural skills that contribute to practical conservation.

The Royal Botanic Garden Serbithang, just outside Bhutan's capital city of Thimpu, was founded in 1999 as an integral part of the National Biodiversity Centre, which also includes the National Herbarium and Seed Bank. Titled 'Institutional Capacity Building and Training', the project ran from 2002 to 2005 and followed on closely from the completion of the *Flora of Bhutan* project which was coordinated and published in Edinburgh from 1975 to 2002. The project was broken down into a series of small sections including staff exchanges in both directions and workshops involving propagation, cultivation techniques, education, interpretation and plant records. The Living Collection at Edinburgh, which is rich in Bhutanese and other Sino-Himalayan plants, was used extensively for practical hands-on training. While this specific project has now come to an end, informal relationships continue, such as staff exchanges between gardens.

The project in Turkey (top left) ran from 2005 to 2008 and was based at Nezahat Gökyiğit Botanik Bahçesi (NGBB), Istanbul and given the title of 'Horticulture and Education for Conservation'. The Garden was established as a park in 1995 and converted to a botanic garden in 2003. This project was also based on a series of staff exchanges and workshops. Horticultural staff from Turkey came to Edinburgh and worked side by side with horticultural staff from RBGE, thereby gaining practical, on-the-job training. Staff from Edinburgh then went to Turkey and led a series of practical workshops in all aspects of plant propagation and cultivation. Other elements of the project included plant records and conservation workshops as well as a demonstration expedition to learn the best techniques of field collecting and recording. Again, the project built on existing collaboration, Edinburgh having coordinated the publication of the *Flora of Turkey* between 1965 and 2000. Again, while the formal part of this project has ceased, the relationship with NGBB is still very strong with training, staff exchanges and joint lecture programmes continuing.

The project in Oman (bottom left and right), which aims to establish a modern national botanic garden to display and champion native species, again came about through RBGE's prior experience in the region, this time with the *Flora of Arabia* project, which has been based at the Garden since the mid 1970s. The difference from the other two projects above is that RBGE was invited to take part at the planning stage, well before any construction work started. RBGE staff worked on the feasibility study, helping to decide which habitats and species should be displayed and what form the interpretation should take. Following the planning phase, the RBGE team helped establish the nursery and habitat displays, working side by side with Omani staff, offering horticultural training and technical support. This programme is still developing and it is envisaged that RBGE will continue to have a strong relationship with Oman for many years to come.

Public engagement

More than 750,000 visits are made to RBGE's four Gardens each year, with visitors attracted by the beautiful, inspiring and secure environments that the Gardens offer. People may come to the Gardens for specific botanical or horticultural purposes, but most just come for a walk, to meet friends and family, feed the pigeons, sunbathe and read the Sunday papers and, of course, to enjoy the plants. These visits offer RBGE an opportunity for education in a much wider sense than the structured education courses.

Through many layers of diverse media, the Garden seeks to engage visitors in the world of plants and the work of RBGE. Whether through interpretation panels, public lectures, guided tours or art exhibits, the essential message is the same: to excite people about plants, their richness and their integral part in our life and culture, and to explain how RBGE works to understand and conserve our plantlife.

RBGE has a long history of public engagement – the Garden was already innovative in Victorian times in having public

above The Garden is for everyone: whether visitors come for in-depth botanical study or simply to enjoy the peace and sunshine, we hope that all who come here absorb something of the importance and beauty of the natural world.

lecture rooms, though the primary focus was on teaching physicians, apothecaries and fellow plantsmen. Guidebooks for the public date back to the 1870s and there have long been maps and leaflets available to orientate and inform visitors, but until very recently the focus was on the horticultural and botanical work rather than public engagement and communication. The transformation into a more public-facing organisation began in the 1980s, with growing awareness of the looming environmental crisis giving a new role and relevance to the Garden's collections and expertise. In response, communication became a far greater priority and resources were increasingly channelled into public outreach and engagement. This change has built momentum ever since, with ever-enhanced maps and signage, publications, shops, talks, events and exhibitions.

Despite these many advancements, at the beginning of the 21st century many regular visitors were unaware that

below Guidebooks dating back to the 1870s tell the story of the Garden's development and demonstrate how public engagement has always been part of RBGE's work.

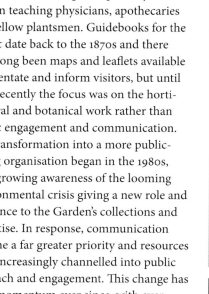

RBGE was a globally renowned scientific institution, or that the Garden was actively engaged in plant conservation in Scotland and around the world. The Edinburgh Garden also lacked any sense of arrival at the West Gate and visitors were confronted with an odd assemblage of old buildings, which had been adapted over the years from stable blocks and a lodge into staff houses, a reception kiosk and a small shop.

The idea of the John Hope Gateway emerged from discussion about the future direction of the Garden and the urgent need for RBGE to explain its purpose more clearly to the public. The vision was realised with the opening of the Gateway in 2009 – a new Edinburgh landmark, equipped to communicate our science and conservation in a

above The John Hope Gateway combines improved visitor facilities with a platform to engage visitors in the work of RBGE and explore the relevance of plants to the critical issues of our time.

right The light-filled foyer of the John Hope Gateway. The building itself embodies the environmental message that the Gateway was designed to communicate, showcasing sustainable technology (such as a wind turbine, biomass boiler, photovoltaics, heat exchangers and rainwater harvesting) and using local materials wherever possible, such as the Caithness stone seen here.

modern, dynamic way. The building includes temporary and permanent exhibition areas, a studio for workshops and demonstrations as well as meeting rooms, a shop and a restaurant with a terrace overlooking the specially created Biodiversity Garden. For the first time in RBGE's history, the quality of the welcome at Edinburgh matches the quality of the Garden itself.

This welcome now extends to all four Gardens. RBGE's Visitor Welcome team recognises that the point of arrival is the most important for enhancing the experience of a Garden visit. All Gardens now have greatly improved facilities, as well as maps and signage and diverse forms of interpretation to tell the Gardens' stories. In 2009, Dawyck became the first 'five star' Garden in Scotland, awarded not only because of the superb garden but also for its new Visitor Centre and the quality of the welcome from staff. Like the Gateway, the Dawyck Visitor Centre has a message about the importance of plants embodied in its building, which uses wood extensively and has a biomass boiler using locally sourced wood and a sedum roof.

Once inside the Gardens, a wide range of media are used to engage visitors. Interpretation is the art and science of explanation and different techniques can be combined to communicate at different levels, including audiovisual displays, information panels, leaflets, guidebooks, tours, lectures and exhibitions. Different approaches are selected as appropriate to the particular

left and below Every plant in the Garden has a story to tell, of its origin, status, collection and cultivation, and interpretation boards bring these stories to life.

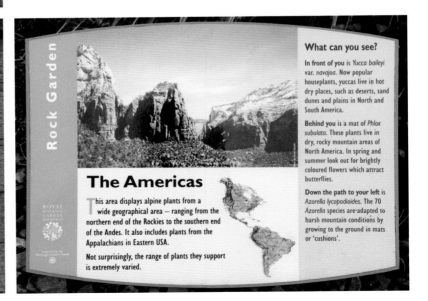

right Visitors can enhance their experience of all four Gardens by hearing plant stories, whether through a self-guided audio trail (top), from our dedicated team of trained volunteer Garden Guides (middle) or from tours by the Curators or Supervisors themselves (bottom).

landscape of the Garden, with a balance sought between the need to share information and a sensitivity to the context in not disrupting the wild atmosphere or view.

For example, in the large space among the towering trees of Dawyck, the interpretation focuses on plant trails that visitors can follow using accompanying maps and leaflets. The David Douglas Trail guides visitors through the Garden to admire the now grand trees planted from seeds collected in the 1820s by the Scottish plant collector David Douglas. Having gazed up at the trees, visitors are encouraged to stop, look down and get acquainted with less conspicuous flora on the Scottish Rare Plants Trail.

The arrangement of the actual plants is an important part of communicating our message and some of the best examples of interpretation are those where the plantings and information have been devised together to tell a particular story. For example, the 'Local Heroes' display at Logan highlights local coastal plants and the Scottish Heath Garden at Edinburgh recreates a native habitat while interpretation panels tell the story of traditional uses of the plants. In both cases, plants in the Living Collection are grouped

together to enable visitors to understand better the plants of Scotland and the issues surrounding their use and conservation.

Guided tours are offered at all Gardens to enable people to engage with the Living Collection in more depth. In Edinburgh and Benmore, there are dedicated teams of volunteer Garden Guides, while at Dawyck and Logan the tours are normally arranged by the Curators and their staff. The tours are adapted according to the season and which plants are looking their best. They cover numerous themes, such as plants collected by a particular explorer or from a geographic region or 'around the world in 80 minutes'. Specialist staff, such as pterid-ologists, mycologists and bryologists, also present tours focusing on ferns, fungi and mosses respectively. 'Dawn Chorus' tours to the birdsong are a highlight of springtime in the Gardens, which are all rich in bird life because of their varied plants and habitats.

A more recent development has been the creation of self-guided audio tours, which use hand-held digital 'wands' to enable visitors to choose the pace at which they explore the Gardens as well as how much detail they wish to hear. The concept was pioneered in 1997 at Logan, an ideal setting for an audio tour as a tool for interpretation, as too many panels or labels would be intrusive in the smaller, more intimate landscape. The audio tours use simple numbered posts beside selected features, plantings or key specimens and the posts can be removed if plants have died or are not flowering at the time – overcoming the practical problem of seasonality in interpretation.

Following their success at Logan, further audio tours have been created for Benmore and for the glasshouses at Edinburgh. There are also special routes available for children with all the audio tours, and the narrative for all ages includes an eclectic mix of first-hand accounts, stories, natural sounds and even songs. The recordings have been compiled from interviews with numerous Garden staff, capturing their enthusiasm for exploring different habitats, discovering new species and caring for the plants back at RBGE.

Art and science come together in the Garden today; art is often used as a vibrant and engaging way to explain scientific principles. Inverleith House, a gallery within the Edinburgh Garden, provides a unique atmosphere in which to host exhibitions of world-class modern art, which always bring new audiences into the Garden. Sometimes these exhibitions provide insights into the Collections, with cultural treasures from RBGE's archives, Library or Herbarium used to enliven the history of the Garden.

All four Gardens now show exhibitions – with Logan's new studio, the Benmore Courtyard Gallery, Dawyck's Studio in its Visitor Centre and Edinburgh's extensive exhibition space in the John Hope Gateway. These are very different spaces but all accommodate touring exhibitions between the Gardens, such as an exhibition of

below Art and science come together at the dedicated exhibition spaces within all four Gardens. RBGE has a long tradition of using art to showcase an appreciation for the natural environment.

above and top right
A Spring Lantern Festival sheds new light on the glasshouses, launching the 2007 'China Now' celebrations and reflecting the long-standing connections between RBGE and China.

bottom right The Palm House illuminated for the 'Power Plant' light show.

photographs of recent fieldwork expeditions to the Himalayas. Art at the Gardens isn't confined to these indoor spaces, however, as recent exhibitions have been held in outdoor Garden areas, using the surrounding plants and landscapes to host pictures of nature that are both beautiful and disturbing. The four Gardens, with their different atmospheres and landscapes, also make ideal venues for outdoor sculpture, particularly when the works are carefully sited to create a unity between sculpture and landscape.

The Edinburgh Garden also becomes a stage in festival time, not only forming a backdrop to many performances but becoming an integral part of them. Many plays by director Toby Gough for the Edinburgh Fringe Festival have drawn from RBGE experiences and help visitors to explore the spaces within the Garden.

Diverse performing arts have enlivened the Garden during the summer season of festivals, including opera, Shakespeare, fairy tales, marching bands, circuses and poetry readings.

Engaging with festivals is an ideal opportunity for RBGE to reach new audiences and work in partnership with other organisations. The Garden takes part in Scottish Biodiversity Week and the National Science and Engineering Week, while Benmore is an important venue for the Cowal Walking Festival. Year-long celebrations, such as the International Year of Biodiversity (2010) and the bicentenary of Charles Darwin's birth (2009), are an increasing influence on exhibitions and events programming.

For many years the Garden has been closely involved with the Edinburgh International Science Festival, offering its own range of lectures, events and

opposite The 'Tree of Hope' invited visitors to respond to the powerful images in Mark Edwards' 2007 exhibition Hard Rain, a photographic commentary on environmental collapse and global poverty. More than 1,000 people took part, sharing their hopes for a fairer world through hand-written messages on colourful ribbons which were tied to the tree.

demonstrations, as well as hosting events for other environmental organisations. These events are based on the Garden's approach to 'Real Life Science', which allows the public to interact with 'real' practitioners of science and horticulture, asking questions, exploring themes and hearing stories from the scientist's real-life adventures – new discoveries, fieldwork in exotic locations or the ways in which plants are used for food and medicine in other cultures.

The informality of presentation and the freedom to ask questions directly to scientists and horticulturists is proving very popular with children and adults alike. For those used to more formal presentations, it can be a revelation to see scientists interact with the public in this way. The style of presentation and ambience of the space allows a rapport and genuine dialogue to develop between the presenter and the public.

This approach – creating a space where people can interact, engage and learn – is at the heart of the Gateway. The Gateway combines all elements of interpretation and informal education, with art, exhibits, interactive displays, lectures, meetings and more.

And so the story comes full circle – the Garden has always responded to the needs of its time and the John Hope Gateway is our response to the needs of the 21st century: enabling visitors to explore the relevance of plants to the critical issues of our time. The central message of the Garden is now communicated through the visitor welcome at all four Gardens and through all our diverse work in public engagement: that plants are fundamental in creating a world that can support all life on this planet, including ourselves.

opposite Art is not confined to RBGE's indoor spaces; the Garden also hosts outside sculpture and travelling exhibitions such as 'Earth from the Air', a breathtaking aerial perspective of our planet's beauty and fragility by French photographer Yann Arthus-Bertrand.

right 'Real Life Science' is at the heart of the Gateway's mission, offering an opportunity for all ages to learn not through lectures but through dynamic engagement and interaction with the real plants and the professional practitioners of science and horticulture.

4
Appendices

Books and journal articles about RBGE and its history

BOWN, D. (1992). *Four Gardens in One, the Royal Botanic Garden Edinburgh.* Edinburgh: HMSO

COWAN, JOHN MCQUEEN (1933). The History of the Royal Botanic Garden, Edinburgh. *Notes R.B.G. Edinb.*, Vol XIX, No XCI, 1–62

COWAN, JOHN MCQUEEN (1935). The History of the Royal Botanic Garden, Edinburgh – the Prestons. *Notes R.B.G. Edinb.*, Vol XIX, No XCII, 63–134

FLETCHER, H.R. & BROWN, W.H. (1970). *The Royal Botanic Garden Edinburgh 1670–1970.* Edinburgh: HMSO

GOVIER, REBECCA ET AL (2001). *Catalogue of Plants 2001.* Edinburgh: Royal Botanic Garden Edinburgh

MATHEW, MANJIL V. (1987). *History of the Royal Botanic Garden Library, Edinburgh.* Edinburgh: HMSO

NOLTIE, HENRY (2011). *John Hope (1725–1786) Alan G. Morton's Memoir of a Scottish Botanist. A new and revised edition.* Edinburgh: Royal Botanic Garden Edinburgh

RAE, DAVID ET AL (2006). *Catalogue of Plants 2006.* Edinburgh: Royal Botanic Garden Edinburgh

RAE, DAVID ET AL (2006). *Collection Policy for the Living Collection.* Edinburgh: Royal Botanic Garden Edinburgh

WATSON, ANDREW (2010). *James Duncan: An Enlightened Victorian.* Edinburgh: Royal Botanic Garden Edinburgh

Books and journal articles about botanic gardens in general, the history of botanic gardens and the history of plant collecting

BROCKWAY, LUCILE H. (1979). *Science and Colonial Expansion, the Role of the British Royal Botanic Gardens.* London: Academic Press

HILL, J. (1915). The history and functions of botanic gardens. *Ann Missouri Bot Gard*, Vol 2, No 1, 185–240

HYAMS, E. & MACQUITTY, W. (1969). *Great Botanical Gardens of the World.* London: Nelson

JUMA, C. (1989). *The Gene Hunters – Biotechnology and the Scramble for Seeds.* London: Zed Books

LINDSAY, ANN (2005). *Seeds of Blood and Beauty, Scotish Plant Explorers.* Edinburgh: Birlinn

MCCRACKEN, D.P. (1997). *Gardens of Empire, Botanical Institutions of the Victorian British Empire.* London and Washington: Leicester University Press

MONEM, NADINE KÄTHE (ED.) (2007). *Botanic Gardens – a Living History.* London: Black Dog Publishing Ltd

MUSGRAVE, TOBY, GARDNER, CHRIS & MUSGRAVE, WILL (1998). *The Plant Hunters, Two Hundred Years of Adventure and Discovery around the World.* London: Ward Lock

OLDFIELD, SARA (2007). *Great Botanic Gardens of the World.* London: New Holland Publishers

OLDFIELD, SARA (2010). *Botanic Gardens, Modern-day Arks.* London: New Holland Publishers

STAFLEU, F.A. (1969). Botanic gardens before 1818. *Boissiera*, No 14, 31–46

Text books and conventions, policies or strategies about conservation relevant to botanic gardens

ANON (1994). *Biodiversity: the UK Action Plan.* London: HMSO

BGCI (1989). *Botanic Gardens Conservation Strategy.* London: BGCI

GUERRANT JR, EDWARD O., HAVENS, KAYRI & MAUNDER, MIKE (EDS) (2006). *Ex situ* Plant Conservation, Supporting species survival in the Wild. Washington: Island Press for the Society for Ecological Restoration International and the Centre for Plant Conservation

HAWKES, J.G., MAXTED, N. & FORD-LLOYD, B.V. (2000). *The Ex Situ Conservation of Plant Genetic Resources.* Dondrecht: Kluwer Academic Press

HEYWOOD, V.H. & WATSON, R.T. (EDS) 1995. *Global Biodiversity Assessment.* Cambridge: Cambridge University Press.

IUCN, UNEP & WWF (1980). *The World Conservation Strategy: Living Resource Conservation for Sustainable Development.* Gland: IUCN

IUCN, UNEP, WWF (1991). *Caring for the Earth.* Gland: IUCN

Botanic garden manuals and booklets

AKEROYD, J.R. & WYSE JACKSON, P. (EDS) (1995). *A Handbook for Botanic Gardens on the Reintroduction of Plants into the Wild*. Richmond: Botanic Gardens Conservation International

CBD SECRETARIAT (2003). *Global Strategy for Plant Conservation*. Montreal: The Secretariat of the Convention on Biological Diversity

DAVIS, K. (2008). *A CBD Manual for Botanic Gardens*. Richmond: Botanic Gardens Conservation International

HAWKINS, B. (2007). *Plants for Life: Medicinal Plant Conservation and Botanic Gardens*. Richmond: Botanic Gardens Conservation International

HAWKINS, B., SHARROCK, S. & HAVENS, K. (2008). *Plants and Climate Change: Which Future?* Richmond: Botanic Gardens Conservation International

LEADLEY, E. AND GREENE, J. (EDS) (1998). *The Darwin Technical Manual for Botanic Gardens*. Richmond: Botanic Gardens Conservation International

OLDFIELD, S. AND MCGOUGH, N. (COMP.) (2007). *A CITES Manual for Botanic Gardens* (2nd edition). Richmond: Botanic Gardens Conservation International

WAYLEN, K. (2006). *Botanic Gardens: Using Biodiversity to Improve Human Well-being*. Richmond: Botanic Gardens Conservation International

WILLIAMS, CHINA, DAVIS, KATE & CHEYNE, PHYLLIDA (2003). *The CBD for Botanists – an Introduction to the Convention on Biological Diversity for People Working with Botanical Collections*. London: Darwin Initiative and Royal Botanic Gardens, Kew

WILLISON, J. (2004). *Education for Sustainable Development: Guidelines for Action in Botanic Gardens*. Richmond: Botanic Gardens Conservation International

WYSE JACKSON, P AND SUTHERLAND, L. (EDS) (2000). *International Agenda for Botanic Gardens in Conservation*. Richmond: Botanic Gardens Conservation International.

Journals and reports produced by RBGE

Annual Report of the Royal Botanic Garden Edinburgh

Edinburgh Journal of Botany

Sibbaldia, The Journal of Botanic Garden Horticulture

The Botanics – a quarterly newsletter

Websites

BOTANIC GARDENS CONSERVATION INTERNATIONAL (BGCI)
www.bgci.org

As well as general topical information about botanic gardens, this site has full details of all the known botanic gardens in the world, a searchable database of plants in botanic gardens and proceedings of botanic garden conferences.

CONVENTION ON BIOLOGICAL DIVERSITY (CBD)
www.cbd.int

This site has a wealth of information about conservation theory and practice as well as details about Conventions and Policies.

PLANTNETWORK – THE PLANT COLLECTIONS NETWORK OF BRITAIN AND IRELAND
www.plantnetwork.org

As well as information about British and Irish botanic gardens this site contains all the presentations made at recent PlantNetwork conferences.

ROYAL BOTANIC GARDEN EDINBURGH
www.rbge.org.uk

UK BIODIVERSITY ACTION PLAN AND SPECIES ACTION PLANS
www.ukbap.org.uk

WORLD CONSERVATION UNION
www.iucn.org

The Living Collection of plants at the Royal Botanic Garden Edinburgh (RBGE) contains in excess of 158,300 individual plants from 157 countries. Table 1 shows the number of families, genera, species, taxa, accessions and plant records arranged by major plant groups. It also shows what percentage of the world's families, genera and species are represented in the Collection.

Table 1. The Living Collection at the RBGE (25 Jan 2011)

Major Taxon	Families		Genera		Species[1]		Taxa[2]	Accessions[3]	Plant records[4]
	No.	% of world	No.	% of world	No.	% of world			
Bryophytes	11	6%	14	1%	14	1%	14	16	16
Fern allies	3	60%	6	60%	57	4%	58	92	143
Ferns	39	100%	162	56%	685	8%	744	1,708	2,556
Gnetophytes	3	100%	3	85%	12	13%	14	29	45
Conifers	8	89%	67	91%	456	68%	891	4,156	17,406
Ginkgophytes	1	100%	1	100%	1	100%	1	12	36
Cycads	4	100%	9	82%	30	23%	30	51	61
Dicots	219	69%	1,801	14%	8,899	3%	11,608	21,654	39,110
Monocots	63	65%	670	18%	3,220	4%	4,022	7,011	9,201
Unknown								47	47
Total	**351**	**50%**	**2,733**	**16%**	**13,374**	**4%**	**17,382**	**34,776**	**68,621**

Notes:

1. A 'species' refers to any unique combination of generic name and specific epithet, even if the actual taxon is of infraspecific rank. Thus, *Vinca major*, *Vinca major* 'Alba' and *Vinca major* ssp. *hirsute* would be counted as one species and three taxa.

2. 'Taxa' includes species, subspecies, varieties, cultivars and hybrids between them.

3. An accession is defined as one or more individuals representing one taxon from one source at one time and in one propagule type (seed, cutting, whole plant etc)

4. A 'plant record' is a subset of an accession, and may represent one or more individual plants, as long those individual plants are planted in one location within the collection.

The figures above record totals for the 'whole' Collection, including plants cultivated offsite as part of the Garden's International Conifer Conservation Programme (ICCP) and plants held in the seed store (genebank).

Where collections are held

RBGE comprises four sites (Inverleith [Edinburgh], Benmore, Dawyck and Logan) that each have widely different climatic characteristics. The Inverleith Garden also has extensive glasshouse collections, nursery facilities and a small seed bank. Table 2 shows the number of plants grown at each of these localities and lists the number of taxa and accessions that are unique to each.

Table 2. RBGE Gardens and their holdings (25 Jan 2011)

Garden	Sub-area	Families	Genera	Species	Taxa		Living accessions	Plant records
					Total	Unique		
Inverleith	Whole site	348	2,628	12,100	14,911	14,536	25,250	36,696
	Glass	314	1,837	6,279	6,873	6,602	11,367	14,301
	Outside	226	1,251	6,419	8,604	7,932	14,183	22,401
	Propagation	217	1,118	3,550	3,940	NA	NA	7,081
Benmore		155	430	1,629	2,451	1,957	5,113	11,841
Dawyck		87	209	1,019	1,241	870	2,212	3,829
Logan		191	671	1,860	2,248	1,561	3,121	4,230
ICCP sites outside RBGE		105	214	554	619	3	2095	11,921

Collection dynamics since last Catalogue

Material 'flows' through the Collection, entering as new accessions from the wild or cultivated sources, and leaving via death or de-accessioning. RBGE receives, on average, just over 2,000 new accessions per year, equating to approximately 8 new accessions entering the Collection every day. Table 3 shows the total accessions, provenance of material and percentage of wild provenance for the last five years.

Table 3. Five-year summary of RBGE accessions

Year	Total accessions	Provenance of material				% wild provenance
		Unknown	Garden (G)	Indirect wild (Z)[1]	Direct wild (W)	
2010	2,149	136	245	140	1,628	83%
2009	2,151	39	306	110	1,696	84%
2008	2,483	16	742	184	1,541	69%
2007	1,926	32	522	154	1,218	71%
2006	1,964	89	551	212	1,112	68%

Notes:

1. 'Indirect wild origin' means material with wild origin details that has been cultivated before being received by RBGE.

Sources of the Living Collection

Table 4 shows provenance numbers and percentages for the whole Living Collection.

Table 4. Wild origin vs garden origin material

Provenance	Living accessions	% of total
Wild origin (W)	18,043	52%
Indirect wild origin (Z)[1]	1,881	5%
Garden origin (G)	13,802	40%
Unknown	1,050	3%
Total	34,776	100%

Notes:

1. 'Indirect wild origin' means material with wild origin details that has been cultivated before being received by RBGE.

Conservation status categories

The World Conservation Union (IUCN) has, for many years, listed criteria for allocating conservation-status categories, or threat categories, to plants (and animals). Table 5 shows the number of plants growing at RBGE under each of the threat categories.

Table 5. Threatened taxa cultivated at RBGE, listed under IUCN categories

UNEP /IUCN 2006 Categories[1]	Taxa at world level	Living taxa at RBGE	Living accessions	Wild collected accessions
Extinct (EX)	95	0	0	0
Extinct in wild (EW)	30	2	4	1
Critically endangered (CR)	1,661	39	327	298
Endangered (EN)	2,413	81	330	262
Vulnerable (VU)	5,068	156	922	711
Data deficient(DD)	745	21	78	58
Conservation dependent (CD)	254	22	72	49
Near threatened (NT)	1,140	85	357	248
Least concern (LC)	1,500	358	2,118	1,303
Total threatened	12,906	764	4,208	2,930

Notes:

1. Categories and their definitions can be found at http://www.iucnredlist.org

Taxonomic specialities of the Living Collection

RBGE has certain taxonomic groups in which it currently specialises or has specialised in the past. Table 6 shows RBGE's top 10 families, ranked in descending order by number of accessions.

Table 6. Plant families with strong representation

| Family | World estimates | | Alive at RBGE | | | | | |
| | Genera | Species | Genera | | Species | | Taxa | Accessions |
			Total	% of world	Total	% of world		
Ericaceae	107	3,400	56	51%	1,007	30%	1,719	4,721
Pinaceae	13	250	11	78%	200	80%	356	1,828
Rosaceae	95	2,825	66	69%	699	24%	913	1,827
Orchidaceae	788	18,500	158	20%	682	3%	785	1,334
Cupressaceae	20	125	31	100%	111	87%	356	1,320
Gesneriaceae	139	2,900	73	52%	409	14%	444	1,028
Compositae	1,528	22,750	159	10%	577	3%	700	914
Iridaceae	82	1,700	32	39%	318	18%	486	889
Liliaceae	288	4,950	14	17%	270	5%	348	730
Sapindaceae	145	1700	14	10%	89	5%	166	390

Geographic specialities of the Living Collection

Just as RBGE specialises in certain plant families, so too it concentrates on certain areas of the world. Table 7 shows RBGE's top 10 countries, ranked in descending order by number of accessions.

Table 7. Major countries of origin for wild collected material

| Country | Estimated number of native species | Alive at RBGE | | | | | |
| | | Families | Genera | Species | | Taxa | Accessions |
		Total	Total	Total	% of country total		
China	32,200	135	366	1,320	4%	1,605	3,300
United States	16,108	107	249	604	4%	657	1,297
Chile	52,842	100	207	391	7%	405	1,195
Japan	5,565	117	243	496	9%	555	1,137
Indonesia	29,375	65	140	373	1%	392	997
United Kingdom	1,623	80	202	396	24%	431	796
Papua New Guinea	11,544	40	98	308	3%	332	758
Nepal	6,973	78	159	341	5%	384	697
New Zealand	2,382	72	134	308	13%	318	606
South Africa	23,420	68	139	362	2%	373	568

III · RBGE REGIUS KEEPERS AND DIRECTORS OF HORTICULTURE

Regius Keepers

Name	Birth and death	Duration as Regius Keeper
James Sutherland	1639–1719	1699–1715
William Arthur	1680–1716	1715
Charles Alston	1685–1760	1716–1760
John Hope	1725–1786	1761–1786
Daniel Rutherford	1749–1819	1786–1819
Robert Graham	1786–1845	1819–1845
John Hutton Balfour	1808–1884	1845–1879
Alexander Dickson	1836–1887	1880–1887
Isaac Bayley Balfour	1853–1922	1888–1922
William Wright Smith	1875–1956	1922–1956
Harold R. Fletcher	1907–1978	1956–1970
Douglas M. Henderson	1927–2007	1970–1987
John McNeill	1933–	1987–1989
David S. Ingram	1941–	1990–1998
Stephen Blackmore	1952–	1999–

Principal Gardeners, Curators and Directors of Horticulture

Name	Birth and death	Duration	Position
James Sutherland	1639–1719	1676	Intendent
John Williamson	Died 1780	c.1756	Principal Gardener
Malcolm McCoig	Died 1789	c.1782	Principal Gardener
Robert Menzies	Died 1800	1789	Principal Gardener
John McKay	1772–1802	1800–1802	Principal Gardener
George Don	c.1764–1814	1802–1806	Principal Gardener
Thomas Sommerville	c.1783–1810	c.1807	Principal Gardener
William McNab	1780–1848	1810–1838	Principal Gardener
James McNab	1810–1878	1849–1878	Principal Gardener
John Sadler	1837–1882	1879–1882	Curator
Robert Lindsay	1846–1913	1883–1896	Curator
Adam Dewar Richardson	1857–1930	1896–1902	Head Gardener
Robert Lewis Harrow	1867–1954	1902–1931	HG 1902, Curator 1924
Lawrence Baxter Stewart	1887–1934	1932–1934	Curator
Roland Edgar Cooper	1891–1962	1934–1950	Curator
Edward E. Kemp	1910–	1950–1971	Curator
Richard (Dick) Shaw	1927–2000	1972–1987	Curator
John Main	1940–	1988–2000	Curator
David Rae	1955–	2001–	Director of Horticulture

V · ACKNOWLEDGEMENTS

Many people have helped in the production of this book and in particular the author would like to thank the people below:

PRODUCTION AND PROJECT MANAGEMENT:
Hamish Adamson

DESIGN: Nye Hughes

EDITING: Anna Levin and Fay Young

COPY EDITING AND PROOFREADING:
Anna Stevenson

INDEXING: Phyllis Van Reenen

REVIEW AND HELP WITH MAJOR SECTIONS OR CHAPTERS: Richard Baines, Peter Baxter, Alan Bennell, Martin Gardner, Mary Gibby, Peter Hollingsworth, David Knott, Leigh Morris, Henry Noltie, Toby Pennington, Graham Stewart

SPECIFIC AND GENERAL INFORMATION AND BOXED CASE STUDIES:
George Argent, Richard Baines, Sadie Barber, Peter Baxter, Stephen Blackmore, Richard Brown, Peter Brownless, David Chamberlain, Laura Cohen, Max Coleman, Jane Corrie, Simon Crutchley, Peter Daniel, Martyn Dickson, Catherine Evans, Natacha Frachon, Louise Galloway, Martin Gardner, Tony Garn, Thomas Gifford, Graham Hardy, David Harris, Michelle Hollingsworth, Kate Hughes, Mark Hughes, Jane Hutcheon, Fiona Inches, Ross Irvine, Susie Kelpie, Greg Kenicer, Catherine Kidner, Sabina Knees, David Long, Phil Lusby, Siobhan McDermott, Amy McDonald, Heather McHaffie, Felicity McKenzie, Elspeth Mackintosh, Michelle Maclaren, Robert Mill, Tony Miller, David Mitchell, John Mitchell, Clare Morter, Paul Nesbitt, Mark Newman, Jacqui Pestell, Leonie Paterson, Alistair Paxton, Axel Dalberg Poulsen, Sally Rae, Jim Ratter, Louis Ronse de Craene, Lesley Scott, Graham Stewart, Alex Twyford, Mark Watson, Emily Wood

PLANT RECORDS INFORMATION:
Rob Cubey, Helen Thompson

SOURCING AND PRODUCING IMAGES:
Richard Baines, Sadie Barber, Peter Baxter, Alan Bennell, Lawrie Buchan, Frieda Christie, Pat Clifford, Kirstin Corrie, David Harris, Elspeth Haston, Shauna Hay, Vlasta Jamnický, Greg Kenicer, David Knott, David Long, Phil Lusby, Neil McCheyne, Amy McDonald, John Mitchell, Leigh Morris, Leonie Paterson, Colin Pendry, Kerstin Price, Steve Scott, Cynthia Shaw, Jane Squirrell, Graham Stewart, Ron Tulloch, Robert Unwin, Donald Wemyss Lynsey Wilson

ADMINISTRATIVE SUPPORT:
Rachel Brown, Kerstin Price

Every effort has been made to trace holders of copyright in text and illustrations. Should there be any inadvertent omissions or errors the publishers will be pleased to correct them for future editions.

IMAGE ACKNOWLEDGEMENTS:

Key Top: (T); Middle: (M); Bottom: (B); Left: (L); Right: (R); Far (F)

Front cover Amy Copeman; p.2/3 Pat Clifford; p.6 Debbie White; p.7 Alexandra Davey; p.8 David Knott; p.10 (L) BBG Staff, (R) David Rae; p.11 (L) David Rae, (R) Lynsey Wilson; p.12 (L & R) David Rae; p.13 (L & R); p.14 (L) David Rae, (R) Pat Clifford; p.15 (L & R) David Rae; p.16 (L) David Knott, (R) David Rae; p.17 (L) Lynsey Wilson, (R) David Rae; p.18/19 Lynsey Wilson; p.20 reproduced with kind permission of the National Library of Scotland; p.21 Nye Hughes; p.24/5 Nye Hughes; p.26/7 reproduced with kind permission of the National Library of Scotland; p.28 (T) Debbie White, (B) reproduced with kind permission of Padua University - Servizio Cerimoniale e Manifestazioni, Università degli Studi di Padova; p.29 reproduced with kind permission of the Royal College of Physicians of Edinburgh; p.30 (L) RBGE Archives, (R) reproduced with kind permission of the Royal Commission on the Ancient and Historical Monuments of Scotland; p.31 RBGE Archives; p.32/3 Helen Pugh/RBGE Library; p.34/5/6 RBGE Archives; p.37 (R) David Rae, (L) Linnean Society; p.38/9 RBGE Archives; p.40 (B) David Rae, (TL & TR) RBGE Archives; p.41 (L & R) RBGE Archives, (M) Linnean Society; p.42/3 RBGE Archives; p.44/5/6/7/8/9 RBGE Archives; page 50 (L) David Rae, (TR & MR & BR) RBGE Archives; p.51 (B) David Rae, (L) Lawrie Buchan, (TR) Ron Tulloch; p.52 (B) David Rae, (T) reproduced with the kind permission of the Royal Caledonian Horticultural Society; p.53/4/5 David Rae; p.56/7/8/9 RBGE Archives; p.60 (TR) RBGE Archives, (L) reproduced with kind permission of the National Portrait Gallery, London, second down (L) RBGE Archives, third down (L) RBGE Archives, (BL) RBGE Archives; p.61 David Rae; p.62 RBGE Archives; p.63 (TL & BL) David Rae, (BR) Debbie White; p.64/5 David Rae; p.66/7 courtesy of David Younger; p.68 (L) courtesy of David Younger, (R) RBGE Archives; p.69/70/1 RBGE Archives; p.72 David Rae; p.73/4 RBGE Archives; p.75 (T) Ron Tulloch, (M) RBGE Archives, (B) Ron Tulloch; p.76 (B) David Rae, (TL & TR) Debbie White; p.77/8 David Rae; p.79 (BL & BR) David Rae, (T) Lynsey Wilson; p.80 David Rae; p.81 Lynsey Wilson; p.82/3 Debbie White; p.84 Peter Baxter; p.85/6 courtesy of David Younger; p.87 Peter Baxter; p.88 (B) David Rae, (T) Neil McCheyne; p.89 (L) David Rae, (R) Peter Baxter; p.90 (TL) David Rae, (TR) Peter Baxter, (BL & BR) Sadie Barber; p.91 Peter Baxter; p.92/3 David Rae; p.94/5 Richard Baines; p.96 Debbie White: p.97 RBGE Archives; p.98 (T & BM) David Rae, (BL) RBGE Archives, (BR) Robert Unwin; p.99 (TL & B) David Rae, (TR) Peter Baxter; p.100 (BL & BR) David Rae, (T) Lynsey Wilson; p.101 David Rae; p.102 (B) David Rae, (T) Lynsey Wilson; p.103 (TL & TR) David Rae, (B) Lynsey Wilson; p.104/5 Lynsey Wilson; p.106 (L) David Knott, (R) David Rae; p.107 (TR) David Binns, (BL) reproduced with kind permission of St Andrews University, (BR) RBGE Archives; p.108 (BR) David Knott, (TL & BL) David Rae; p.109 David Knott; p.110 (T) Lynsey Wilson, (B) Peter Baxter; p.111 (T) David Knott, (B) David Rae; p.112 David Knott; p.113 (RB) David Rae, (RT) Graham Stewart, (T) Mark Watson; p.114 Nye Hughes; p.116/17 David Rae; p.118 (TL) David Rae, (TR) Greg Kenicer; p.119 (TL) David Knott, (TR & BL & BR) David Rae; p.120/1 David Rae; p.123 (TL) Colin Pendry, (BL) Mark Watson, (R) Peter Hollingsworth; p.124 David Rae; p.125 (BL) Bill McNamara, (T)David Rae, (BR) Matsushita Hirotaka; p.126 (TL) CBD, (TR) Colin Pendry; p.127 (TR) David Knott, (TL & B) Sadie Barber; p.128 (TL & BR) Ben Dell, (TR) Colin Pendry, (BL) David Rae; p.129 Tony Miller; p.130 Steve Scott; p.131 (TL & BL & BM & BR) David Rae, (TR) Mark Watson; p.132 David Rae; p.133 (L) David Knott, (TR) David Rae, (MR & BR) Mark Watson; p.134/6/7 Sadie Barber; p.138 (TL & BL) Pat Clifford, (R) Robert Unwin; p.139 Robert Unwin; p.140 David Rae; p.141 Robert Unwin; p.142 David Rae; p.143 (T) Debbie White, (ML) Peter Gosling, (BL) Peter Gosling, (BR) Tom Christian; p.144 (TR) David Rae, (BR) John Mitchell, (L) Sadie Barber; p.145 (R) David Rae, (TL) Natacha Frachon, (BL) Sadie Barber; p.146 David Rae; p.147 (R) David Knott, (L) Sadie Barber; p.148 David Rae; p.149 row 1 (L & M) David Rae, row 2 (L & M & R) David Rae, row 3 (L & M) David Rae, row 4 (L & M & R) David Rae, row 1 (R) Richard Baines, row 3 (R) Richard Baines; p.150 Neil McCheyne; p.151 (R) David Rae, (L) Lynsey Wilson; p.152 row 1 (L & R)Neil McCheyne, row 2 (L & R) David Rae, row 3 (L & R) Sadie Barber, row 4 (L & R) left Neil McCheyne; p.153 row 1 (R) Neil McCheyne, row 2 (L & R)Neil McCheyne, row 3 (L & R) Neil McCheyne, row 4 (R) Neil McCheyne, row 1 (L) Sadie Barber; p.154 (BR) Alistair Paxton, (L & TR) Pat Clifford; p.155 Sadie Barber; p.156/7 Debbie White; p.158/9 Lynsey Wilson; p.160 Rob Cubey; p.161 Debbie White; p.163/4 David Rae; p.165 Sadie Barber; p.166 Natacha Frachon; p.167 (M & B) Lynsey Wilson, (T) Mark Watson; p.169 Helen Pugh/RBGE Library; p.172/3 Nye Hughes; p.174/5 Antje Majewski; p.176 (BL & R) Lynsey Wilson, (TL) Scottish Montane Willow Group; p.177 (TL & BL & BR) David Rae, (TR) Robert Unwin; p.178 (TL) David Rae, (TR) Paulina Hechenleitner; p.180 (T) Debbie White, SEM Image (L & R) Frieda Christie, (B) Leigh Morris; p.181 (B) David Long, (T & M) David Rae; p.182 (TL) David Harris, (BL & BR) Lynsey Wilson, (TR) Martin F. Gardner; p.183 Michelle Hollingsworth; p.184 (TL & TR) David Long, (M) Michelle Hollingsworth; p.185 David Long; p.186 Axel Dalberg Poulsen; p.187 (TL & TM & TR) Mark Hughes, (MR & BL) Michael Moeller, (BR) reproduced with kind permission of American Society of Plant Biologists; p.188/9 David Rae, (BR) BGCI; p.190 (BR) David Rae, (TL & TR) Martin F. Gardner, (BL) Nye Hughes; p.191 (MR & BR) David Rae, (TR) Ian Edwards; p.192 (TM) David Rae, (T) Debbie White, (BL) Kate Hughes, (BR) Tom Christian; p.193 Jane Squirrell; p.194 (LM & RM & LB) Natacha Frachon, (TL & RM & RB & R) Scottish Montane Willow Group; p.195 (BL) Graham French, (BM & BR) Deborah Kohn, (T) Jane Squirrell; p.196 Stuart Lindsay; p.197 Debbie White; p.198 (TM) David Rae, (BM & BR) Heather McHaffie, (TR & BL) Natacha Frachon, (TL) Sadie Barber; p.199 (BL) Lynsey Wilson, (TR & MR) Sadie Barber; p.200 (TL) Antonio Lara Aguilar, (R) David Rae, (BL) Heather McHaffie; p.201 Sadie Barber; p.202 (BL) Natacha Frachon, (BR) Neil Stuart; p.203 (T) Kerry S. Walter, (BL) Lisa Banfield, (BR) Steve Scott; p.204/5 Brenda White; p.206 RBGE Archives; p.207 (TR & BL & BM & BR) Brenda White, (TL) Debbie White; p.208 Brenda White; p.209 (TM) Brenda White, (TL & TR & B) Cath Evans; p.210 Debbie White; p.211 (TR & BR) Leigh Morris, (TL & ML & BL) RBGE Archives; p.212 (M & BL) Brenda White, (TL) Tim Rich, second row (L) Rhiannon Crichton, (TR) David Rae; p.213 (B) Brenda White, (T) Debbie White; p.214 (TL) Jacqui Pestell, (BL) Leigh Morris, (BR) Sadie Barber; p.215 David Rae; p.216 Lynsey Wilson; p.217 Matt Laver; p.218 David Rae; p.219 (B) Debbie White, (M) Tony Garn, (T) Vlasta Jamnický; p.220 (TL & BL & BR) Chris Watt, (TR) Kirsty Bennell; p.221 (BL) Paul Nesbitt/Courtesy of the artist and Michael Werner Gallery, New York, (R & T) Lynsey Wilson, (BM) Paul Nesbitt/Courtesy of the artist and the Eggleston Artistic Trust; p.222 (BR) Vlasta Jamnický, (L & TR) Sadie Barber; p.223 Vlasta Jamnický; p.224 Debbie White; p.225 (B) Lynsey Wilson, (T) Max Coleman; p.226/7 Nye Hughes.